ENERGY TRANSITION AND CARBON NEUTRALITY IN ASEAN

DEVELOPING CARBON CAPTURE, UTILIZATION AND STORAGE TECHNOLOGIES

Series in Energy Transition, Carbon Neutrality, and Sustainability

Print ISSN: 2972-3957
Online ISSN: 2972-3965

Series Editor: Farhad Taghizadeh-Hesary *(Tokai University, Japan)*

Published

Series in Energy Transition, Carbon Neutrality, and Sustainability - Volume 2

ENERGY TRANSITION AND CARBON NEUTRALITY IN ASEAN
DEVELOPING CARBON CAPTURE, UTILIZATION AND STORAGE TECHNOLOGIES

Edited by

Han Phoumin
*Economic Research Institute for ASEAN, and East Asia
(ERIA), Indonesia*

Rabindra Nepal
University of Wollongong, Australia

World Scientific

NEW JERSEY · LONDON · SINGAPORE · BEIJING · SHANGHAI · HONG KONG · TAIPEI · CHENNAI · TOKYO

Published by

World Scientific Publishing Co. Pte. Ltd.
5 Toh Tuck Link, Singapore 596224
USA office: 27 Warren Street, Suite 401-402, Hackensack, NJ 07601
UK office: 57 Shelton Street, Covent Garden, London WC2H 9HE

Library of Congress Cataloging-in-Publication Data
Names: Han, Phoumin, editor. | Nepal, Rabindra, editor.
Title: Energy transition and carbon neutrality in ASEAN : developing carbon capture, utilization
    and storage technologies / editor, Phoumin Han, Economic Research Institute for ASEAN and
    East-Asia, Indonesia, Rabindra Nepal, University of Wollongong, Australia.
Description: New Jersey : World Scientific, [2025] | Series: Series in energy transition, carbon
    neutrality, and sustainability, 2972-3957 ; vol. 2 | Includes bibliographical references and index.
Identifiers: LCCN 2024018801 | ISBN 9789811288043 (hardcover) |
    ISBN 9789811288050 (ebook) | ISBN 9789811288067 (ebook other)
Subjects: LCSH: Carbon sequestration--Southeast Asia. | Carbon dioxide mitigation--
    Southeast Asia. | Energy development--Environmental aspects--Southeast Asia. |
    Energy policy--Environmental aspects--Southeast Asia.
Classification: LCC TD885.5.C3 E53435 2025 | DDC 333.791/50959--dc23/eng/20240606
LC record available at https://lccn.loc.gov/2024018801

British Library Cataloguing-in-Publication Data
A catalogue record for this book is available from the British Library.

For any available supplementary material, please visit
https://www.worldscientific.com/worldscibooks/10.1142/13725#t=suppl

Desk Editors: Aanand Jayaraman/Sandhya Venkatesh

Typeset by Stallion Press
Email: enquiries@stallionpress.com

About the Editors

Han Phoumin is Senior Energy Economist with the Economic Research Institute for ASEAN and East Asia (ERIA). Han specializes in economic development and policy and applied econometrics, drawing on more than 20 years of experience at various international and intergovernmental organizations and multidisciplinary research consortiums related to energy markets and technologies, the environment, integrated water resource management, governance, and economic development in the Association of Southeast Asian Nations (ASEAN) and East Asia. Much of his career over the past decade has involved the energy sector, including sustainable hydropower development, renewable research, efficiency, clean coal, security, and supply and demand forecasting.

Rabindra Nepal is Associate Professor of Economics, School of Business, Faculty of Business and Law, University of Wollongong, Australia. An internationally recognized economist specializing in economic and policy analysis concerning energy, environment, and resources, Nepal is widely published, with more than 100 journal articles and book chapters, in addition to three co-edited volumes. With a strong policy focus in his research, he edits international journals with a similar emphasis, including serving as co-editor of the journal *Economic Analysis and Policy* and being appointed to the International Advisory Board of the journal *Energy Policy*.

Acknowledgments

This book is the product of a working group of the Economic Research Institute for ASEAN and East Asia (ERIA) which carried out a study under the auspices of ERIA's energy research section on low-carbon emission technologies including carbon capture, utilization, and storage (CCUS). The researchers selected for this study group explore ways in which CCUS applications, practices, and experiences can be shared among stakeholders to ensure their rapid commercialization under various business models. We hereby thank all participants for their support, including the interviewees who provided country-specific data and other information. Special thanks to Prof. Tetsuya Watanabe, President, and Prof. Hidetoshi Nishimura, former President, of ERIA for their advice and encouragement, which ensured that this project would address contemporary energy issues in ASEAN and East Asia. We also thank Michael House, Word House Ltd., Tokyo, Japan, for copyediting the text. Last but not least, we are grateful to Amanda Yun of World Scientific for managing the publishing process.

Contents

Introduction

One technological pathway for the global energy sector to achieve carbon neutrality and net-zero CO_2 emissions by 2050 is by developing carbon, capture, utilization, and storage (CCUS). CCUS can improve air quality by extracting CO_2 emitted into the atmosphere by energy generation facilities that burn carbon-intensive fossil fuels. According to the 2023 International Energy Agency (IEA) report on tracking clean energy progress, global CCUS facilities currently capture more than 45 $mtCO_2e$ annually. The report adds that the United States, European Union, and United Kingdom have made notable progress in CCUS adoption, and CCUS projects have also come online in China and Japan.

Nonetheless, global CCUS development overall lags well behind what is required to achieve net-zero, which is particularly concerning for the Association of Southeast Asian Nations (ASEAN), as this region is projected to remain dependent on fossil fuels, particularly coal, for electricity generation, while demand is projected to triple by 2050. At the same time, all ASEAN member states except the Philippines and Indonesia have also pledged to achieve net-zero by 2050. This makes CCUS vital to ASEAN member states' decarbonization efforts, emphasized by the fact Indonesia alone has instead set a net-zero target date of 2060 and established a CCUS regulatory framework.

The problem is that significant barriers persist to developing CCUS in ASEAN. One is that relevant policies and regulatory frameworks still remain a work in progress in many ASEAN member states. Another is the need for large and risky investments amid concerns about readiness and public acceptability that complicate efforts to encourage public–private CCUS partnerships. Possible solutions include innovations that reduce capture costs, such as moving away from chemical absorption and separation and toward membranes and looping.

CCUS deployment in ASEAN is thus clearly neither a matter of quick policy fixes nor technological remedies. Rather, an integrated approach to understanding the policy, technological, and economic aspects of CCUS deployment is necessary, given that CCUS deployment is still in the exploratory stages in many ASEAN member states, as indicated earlier. The aforementioned regulatory framework, adequate financial support, investigating political economy and encouraging social acceptability, and strengthening regional CCUS infrastructure cooperation will take policy precedence initially.

This book is a response to these concerns. It offers 11 policy-driven empirical studies investigating and evaluating facets of CCUS development and deployment in ASEAN, organized into three parts. Part 1 constituting 4 chapters maps CCS/CCUS in ASEAN's regional and country-specific policy frameworks.

In Chapter 1, Phoumin, Kimura, and Nepal apply energy modeling techniques to analyze possible pathways for a low-carbon energy transition that would assist ASEAN and East Asia in reaching carbon neutrality by 2050. To this end, the authors evaluate CCS/CCUS applications in the energy sector in the East Asia Summit (EAS) region and provide a general policy framework for affordable deployment. This chapter sheds light on decision-making policy implications for ensuring that the region enjoys economic growth and investment opportunities without compromising energy security or the environment.

In Chapter 2, Setyawati assesses the future of ASEAN decarbonization through CCUS deployment from an energy justice perspective. This novel application analyzes the social and economic aspects of energy systems to ensure fair allocation of energy provision benefits and responsibilities, based on the current 10 projects that are underway in ASEAN. Results show the potential for injustice in CCUS development and call on policymakers to embrace the eight principles of energy justice in their plans.

Chapter 3 by Chantanakome is a study of how Thailand grapples with rising energy demands and greenhouse gas emissions (GHGs) and an attempt at understanding the role of CCUS in decarbonization. It concludes that a successful Thai CCUS deployment requires collaboration between government and industry to establish a sustainable market and reduce costs.

In Chapter 4, Ahmad *et al.* aim to comprehend the global CCUS ecosystem, from which they draw lessons for ASEAN by reviewing other countries' CCUS frameworks and focusing on the exploratory role of cleantech startups. They recommend encouraging expert CCUS training and that ASEAN member states create a dedicated fund for developing and deploying CCUS through domestic startups.

Part 2 comprising 3 chapters captures technological aspects of CCUS deployment by focusing on existing and potential industrial applications.

In Chapter 5, Sasaki and Vijitharan examine nature-based CCUS, focusing on the carbon capture potential and timber production outcomes of conventional and reduced impact logging in the Mekong region from 1990 to 2050. Implementing forest management in the region has the potential, contingent upon prevailing carbon pricing mechanisms, to generate significant carbon revenues, estimated at USD 3500.07 million across the 2020–2030 Paris Agreement period. Findings advocate adopting sustainable forest management practices in the Mekong region to alleviate pressure on natural forests, ensure sustained timber production, foster enhanced ecosystem services, and effectively combat climate change.

In Chapter 6, Hardiansyah *et al.* explore the potential of membranes to reduce carbon capture costs, made possible by a new generation of membranes made from two-dimensional materials, such as graphene, with improved carbonaceous gas separation and selectivity. The structures were evaluated by advanced material characterization for performance against gaseous compounds. Findings show that membranes can be easily implemented with supportive government policy and infrastructure to fulfill ASEAN energy demands and carbon neutrality targets.

In Chapter 7, Sasaki provides a comprehensive review of past efficiencies in harvesting, processing, and manufacturing along the Southeast Asian wood industry supply chain. This matters because the global wood industry annually harvests and consumes approximately 2.2 billion cubic meters (Gm^3) of wood, presenting an opportunity for greater carbon storage in harvested wood products through strategic implementation of Industry 4.0 technologies. Findings advocate for sustainable practices aligned with global climate goals and promoting a greener future by addressing the Industry 4.0 knowledge gap.

Part 3 includes 4 chapters and focuses on the economic aspects of CCUS development and deployment.

In Chapter 8, Dutta *et al.* investigate Chinese carbon market volatility, acknowledging that rising economic uncertainty, geopolitical conflicts, and oil price volatility impede CCUS deployment despite its importance in mitigating climate change. Findings show that time-dependent jumps occur in this market, which the aforementioned satisfactorily account for.

Chapter 9, by Dong *et al.*, investigates the impact of CCUS innovation on green total factor productivity based on panel data for 2007–2019. Findings show a significant positive impact on efficiency in industrial infrastructure and carbon emissions. The researchers offer practical policy suggestions to encourage these outcomes.

Chapter 10, by Taghizadeh-Hesary *et al.*, provides practical policy recommendations for making CCUS affordable by reviewing the literature and corporate policies implemented to date, concluding that technological advancement, policy support, and market mechanisms are required.

In Chapter 11, Havyatt *et al.* analyze CCUS in political economy terms for its potential to dramatically address global climate change. The researchers adopt a frame analysis tool for this purpose, synthesizing five key propositions, concluding that inherent political economy challenges to CCUS must be overcome if the strategy of overtly relying on its eventual development to combat climate change is to succeed.

Part 1

Chapter 1

Carbon Neutrality Pathways in East Asia: A Consideration of CCUS in the Power Generation Sector

Han Phoumin[*,‡], Shigeru Kimura[*,§], and Rabindra Nepal[†,¶]

*Economic Research Institute for ASEAN and East Asia,
Jakarta, Indonesia*

†*School of Business, Faculty of Business and Law,
University of Wollongong, Sydney, Australia*

‡*han.phoumin@eria.org*

§*shigeru.kimura@eria.org*

¶*nepal@uow.edu.au*

Abstract

While countries pledged net-zero emissions by mid-century at COP 26 and COP 27, ongoing oil and gas price hikes arising from Russia's invasion of Ukraine may discourage switching from coal to natural gas, which is a low-hanging mitigation opportunity for fossil fuel-dependent regions. Fossil fuels, especially coal, could thus remain in use longer than previously anticipated in some Asian countries' energy mixes. This chapter explores the low-carbon energy transition (LCET) scenario by way of analyzing the impact of net-zero emission (NZE) technologies that encourage achieving carbon neutrality as early as 2050. It evaluates such clean technologies as carbon capture, utilization, and storage (CCUS) in energy generation, alongside renewables and other clean fuels. Energy modeling techniques suggest how East Asia might best meet its carbon-neutrality pledges. Our findings shed light on policy implications for ensuring that East Asia enjoys economic growth and investment opportunities while avoiding threats to energy security and environmental hazards. It suggests there may be multiple low-carbon energy transition pathways that countries may take to reach carbon neutrality by 2050.

Keywords: Carbon neutrality, renewable energy, clean fuels, and clean technologies.

1. Introduction

The Economic Research Institute for ASEAN and East Asia (ERIA) issues biennial updates to its energy outlook and conservation potential analysis for the East Asia Summit (EAS) region, most recently as of this writing in 2022–2024.

Global concerns about rising global oil and gas prices have escalated since February 24, 2022, when Russia invaded Ukraine. A barrel of Brent Crude was USD 95.42 on February 24, reaching USD 127.98 on March 8, falling to USD 87 on November 18 (OilPrice.com, 2022). Natural gas, the price of which is indexed to the global oil price, has temporarily risen even higher, with East Asia importing LNG to meet demand. Hence, the Japan/Korean Marker (JKM (Platts)) price fell slightly from USD 32.47/MBTU in September 2022 to USD 27/MBTU on November 18, 2022. So long as Russia's war on Ukraine continues, however, and no other fossil fuel supplies replace what the fighting there has taken off the market, the aforementioned sentiment is likely to persist.

Another consequence of Russia's ongoing aggression against Ukraine is that, despite countries pledging net-zero emissions by mid-century at COP 26 and COP 27, the aforementioned fossil fuel price hikes may discourage switching from coal to natural gas, a low-hanging mitigation opportunity for fossil fuel-dependent regions. Even when all nationally determined contributions (NDCs) are revised and implemented, coal, oil, and natural gas will still occupy over half of the total global primary energy supply. This is particularly true of Asia, which comprises the bulk of incremental energy demand and CO_2 emissions, making clean use of fossil fuels and carbon capture, utilization, and storage (CCUS) crucial to decarbonization.

A recent ERIA study on the Association of Southeast Asian Nations' (ASEAN) decarbonization scenarios shows that the marginal abatement cost for achieving net-zero emissions by 2050 will be much higher than for net zero by 2070 (Kimura *et al.*, 2022). Although some countries have yet to pledge any specific targets for carbon neutrality in their NDCs, many developing nations, in addition to developed ones, are already fully committed to making the energy transition from fossil fuel-based energy systems to greener energy systems in which renewable energies such as wind, solar, hydropower, and biomass will have a larger share in energy mixes, along with other clean fuels such as hydrogen and ammonia (Akira and Han, 2022). CCUS will be required to manage the remaining emissions, including those from energy and other industries that use fossil fuels, thereby also advancing energy security goals.

CCUS is crucial to clean energy transitions, as it is the only technology that contributes both to directly reducing emissions in energy-intensive economic sectors and to removing CO_2 to balance unavoidable emissions. Captured CO_2 can be utilized, rather than sequestered, to make everything from fuel, concrete, and shoes to cleaning products, plastics, and food. This allows us to recycle emissions and create a circular carbon economy. CCUS deployment will, however, require appropriate supporting policies and investment, in turn requiring that experts understand the technologies sufficiently to help decision-makers institute appropriately supportive CCUS policies.

Generally, there is no one-size-fits-all approach in terms of decarbonization pathways. It will need to consider socioeconomic and political circumstances capable of assisting countries in reaching carbon neutrality. Accordingly, this chapter will explore the low-carbon energy transition (LCET) carbon-neutral scenario as regards the potential of CCUS in energy sector decarbonization.

1.1. *East Asia Summit*

The East Asia Summit (EAS17) is a collection of diverse countries in terms of per capita income and energy consumption, standards of living, resources, and climate. These include the ASEAN member states of Brunei Darussalam, Cambodia, Indonesia, Lao People's Democratic Republic (Lao PDR), Malaysia, Myanmar, the Philippines, Singapore, Thailand, and Vietnam, as well as Australia, China, India, Japan, Republic of Korea (South Korea), New Zealand, and the US.

Whereas some EAS17 members are developed economies, the majority are developing states. Several had per capita gross domestic product (GDP) less than USD 1,500[1] in 2019, while others had more than USD 53,000. Developed countries also tend to have higher energy consumption per capita than developing countries. In fact, large shares of the populations of the developing states still meet their energy needs chiefly with traditional biomass fuels.

These differences partly explain why energy efficiency and conservation (EEC) goals, action plans, and policies are assigned different priorities across countries. Developed economies may be eager to reduce energy consumption, whereas developing countries tend to emphasize economic growth and improving their standards of living, which typically result in their own increased energy consumption per capita as well.

1.2. *Objective and rationale*

This study aims to analyze the potential impact of proposed additional EAS17 energy-saving goals, action plans, and policies on energy consumption and GHG emissions by fuel and sector. It also explores the potential for CCUS decarbonization in the energy sector and provides a platform for energy collaboration and capacity building on energy modeling and policy among member states.

The study supports the Cebu Declaration (ASEAN Secretariat, 2007), which specifies such goals as the following:

- improving fossil fuel efficiency and environmental impact;
- reducing dependence on conventional fuels through intensified EEC programs, increased share of hydropower, expansion of renewables and biofuel production/utilization, and, for interested parties, commercial nuclear power; and

---

[1]All USD figures herein are at constant year 2015 values unless otherwise specified.

- contributing to climate change abatement by mitigating GHG emissions through policies including low-carbon technologies such as CCS and CCUS.

The Government of Japan, in its capacity as the coordinator of the energy efficiency work stream of the Energy Cooperation Task Force (ECTF), asked ERIA to investigate energy conservation and CO_2 emissions reduction potential in East Asia. ERIA accordingly convened a Working Group on the Analysis of Energy Savings Potential, with participants from all EAS17 member states.

1.3. *Enabling CCUS deployment in Asian*

The 2022 G20 Indonesian Presidency Energy Transition statement has as its primary subject "Global Cleaner Energy Systems and Just Transitions," with the purpose of "achieving global deal to accelerate energy transition" (G20 Indonesia, 2022). Multiple pathways will be required to reach this goal. One such is CCUS, which contributes by reducing emissions from existing energy assets and hard-to-abate sectors, providing a cost-effective pathway for low-carbon hydrogen production, and removing carbon from the atmosphere. It is accordingly urgent that G20 members comprehend possible low-carbon energy transition pathways while implementing the 2021 G20 Italian action plan that includes ensuring access to clean energy for all, especially in developing countries (ERIA, 2022a).

Although there is consensus on the need to reduce global warming, the means to that end vary widely according to circumstances, especially as many countries in the G20 and the developing world alike are still highly dependent on fossil fuels. COP 26 is accordingly influencing nations the world over to become low-carbon societies to limit global warming to below 2°C above pre-industrial levels, if not below 1.5°C (United Nations, 2023). Among the decisions agreed to at COP 26 are improved efforts to build resilience to climate change, curb GHGs, and provide necessary finance for both. While all countries must reduce emissions, it is clearly a great challenge for developing countries with high carbon intensities to achieve carbon neutrality, given the impact this will have on energy affordability. This is particularly true in Southeast Asia, where 90 percent of energy demand growth since 2000 has been driven by fossil fuels. Therefore, the developing world must necessarily rely on technologies such as CCUS to decarbonize emissions. And reducing costs and increasing CCUS deployment in the developing world is necessarily key to this transition (ERIA's Press Release, 2023).

Even if all NDCs are implemented, coal, oil, and natural gas will still occupy over half of the total global primary energy supply (Kimura *et al.*, 2023). This is particularly true in Asia, which comprises the bulk of incremental energy demand and CO_2 emissions due to being highly dependent on fossil fuels. Clean fossil fuel use and CCUS will thus be crucial to decarbonizing emissions.

Interest in CCUS is growing the world over in response to rising ambitions to fight climate change. In 2021 and 2022, plans for the first CCS facilities were enacted in countries including Indonesia, Malaysia, and Thailand. Applications are also becoming more diverse, with projects applying CCUS to such sectors as natural gas liquefaction and cement (ERIA, 2022). The first commercial direct air capture (DAC) with storage project was also recently commissioned in Europe. In East Asia, Japan strives to develop CCUS through such demonstrations as the Tomakomai Project. CCUS could offer leeway to Indonesia and other countries around the world striving for net-zero emissions, given their current dependence on fossil fuel power plants. Indonesia's CCUS initiatives have been started by key energy players. An equatorial belt that serves as a hub for CCUS could be an interesting concept to explore, as it could encourage integration and collaboration among government, developers, and financiers.

A critical next step for Indonesia is to launch a pilot project that could lead to a commercial-scale enhanced oil recovery (EOR) project that could store 14 $GtCO_2$ over 15 years and increase the oil recovery rate therein by 14 percent. As with many Southeast Asian countries, CCUS in Indonesia involves numerous agencies, ministries, and sectors, and understanding the roles of all parties is essential to forging fruitful collaborations (ERIA, 2022). Pertamina takes a multifaceted approach to unlocking CCUS, partaking in selecting injection sites, initiating collaborations for feasibility studies, and exploring CO_2 utilization possibilities. The company has received wide-ranging support from Indonesia's Ministry of Education, Culture, Research, and Technology, the Indonesia Endowment Funds for Education, and the Ministry of Finance. Experts agree that technological advancements can reduce capital expenditures (CAPEX) and operational costs, making CCUS more affordable.

2. Data and Methodology

2.1. *Scenarios*

This study examines the following scenarios, pursuant to other studies conducted annually from 2007 to the present. The first is business as usual (BAU), reflecting each country's current goals, action plans, and policies. The second is an alternative policy scenario (APS), including additional goals, action plans, and policies, per annual reports to the East Asia Energy Ministers Meeting (EAS–EMM). The APS assumptions are as follows: more efficient energy consumption (APS1), more efficient fossil fuel energy generation (APS2), more consumption of new and renewable energy (NRE) and biofuels (APS3), and introduction or greater utilization of nuclear energy (APS4). There is a temptation in the LCET to combine the APS with additional options including such clean fuels as hydrogen and ammonia and technologies such as CCUS. The energy models can estimate the individual impacts of these assumptions on both primary energy supply and CO_2 emissions. While the main report highlights only the main scenarios of BAU, APS, and LCET, all of the APS scenarios are analyzed by the country as well.

Detailed APS assumptions are as follows:

- APS1 assumes energy consumption reduction targets by sector, use of more efficient technologies, and energy-saving practices in the industrial, transport, residential, commercial, and agricultural sectors for some countries. This scenario results in fewer CO_2 emissions in proportion to energy consumption reductions.
- APS2 assumes the energy sector utilizes more efficient thermal power plant technologies, resulting in lower inputs and CO_2 emissions in proportion to efficiency improvements. The most efficient coal and natural gas combined-cycle technologies are assumed to be utilized for new power plant construction in this scenario.
- APS3 assumes higher NRE contributions for electricity generation and utilization of liquid biofuels in transportation, resulting in lower CO_2 emissions as NRE is considered carbon neutral. Inputs may not decrease, however, as NRE, like biomass and geothermal energy, is assumed to be less efficient than fossil fuel-fired generation when converting electricity generated on the basis of input.

- APS4 assumes either nuclear energy adoption or increased use in countries already using it. This scenario would produce fewer CO_2 emissions as nuclear energy emits minimal CO_2 because nuclear thermal conversion efficiency is assumed at 33 percent, on par with BAU.

Detailed LCET assumptions are that under APS, which include more progress in switching from fossil fuels to hydrogen, electricity, and biomass in transport and industry, as well as CCUS in industry and energy.

- Switching from coal to combined cycle gas turbine (CCGT) is considered transitional. Starting in 2035, hydrogen will be introduced in industry replacing coal in iron and steel, and diesel in other sectors, reaching 100 percent utilization by 2050. Hydrogen and ammonia use in industry will begin after 2040, including co-firing in energy and boilers. Biomass will also replace coal and natural gas in other sectors in select countries beginning in 2030, reaching 95 percent utilization by 2050.
- Electric vehicles (EV) will replace diesel and gasoline in public transport by 2035, and as many as 70 percent of privately owned vehicles will be electric in select countries between 2025 and 2050.
- CCUS will be applied to coal and natural gas-fired power in cement and energy beginning in 2040, reaching 100 percent utilization by 2050.

The foregoing assumptions stipulate the aforementioned differences among EAS17 member states in terms of progress to date in implementing EEC goals, action plans, and policies. While some countries already have significant energy-saving goals, action plans, and policies built into BAU, others have only started.

2.2. *Data*

For consistency, the historical energy data used in this analysis are converted from International Energy Agency (IEA) energy balance tables to the energy balance tables of EAS17 member states' national energy statistics. For ASEAN member states, Cambodia, Lao PDR, and Myanmar, comprising the CLM, use their own national energy statistics produced with ERIA support, while the remainder use the APEC energy database, which includes their national energy data. The other EAS member states,

China and India, use the aforementioned IEA historical energy balance tables (IEA, 2020b). Socioeconomic data were obtained from the World Bank's online World Databank, World Development Indicators (WDI), and Global Development Finance. Other data, including transportation, buildings, and industrial indices, were provided by Working Group members where available. Where official data are not available, estimates are obtained from other sources or developed by the Institute of Energy Economics, Japan (IEEJ), especially regarding crude oil and other international energy prices.

2.3. *Methodology*

In 2007, the primary model used was IEEJ's World Energy Outlook Model, which was also used in preparing the Asia/World Energy Outlook. Beginning in 2008, ASEAN member states used their own energy models, while other EAS member states were still following the IEEJ model, providing their own key assumptions on population. The following is a brief description of these models.

ASEAN member states' energy models were developed by applying econometrics to forecasts of energy balance tables based on consumption and input/output in energy generation. Consumption is forecast with demand equations by energy and sector and future macroeconomic assumptions. For this study, all countries used the LEAP software.

IEEJ produced energy outlooks of other countries using its model, with explanatory variables based on exogenously specified GDP growth rates. The IEEJ model similarly projects natural gas and coal prices based on exogenously specified oil price assumptions. Demand equations are econometrically calculated in another module using historical data, with future parameters projected using the explanatory variables. Using econometrics means that, while historical trends will influence future supply and demand, energy supply and new technologies are treated exogenously. For electricity generation, the Working Group was asked to specify assumptions about member states' future electricity generation mixes by source.

3. Assumptions

Growth in energy consumption and GHGs in EAS17 as elsewhere is driven by such socioeconomic factors as population and economic growth as well as increasing vehicle ownership and electricity access. Together, they create what might be called a growth headwind that obstructs efforts to reduce consumption. While this headwind must be understood as part of any regional energy demand analysis, increased energy consumption remains essential to achieving socioeconomic development goals. This section discusses assumptions about key EAS17 socioeconomic indicators and energy policies between now and 2050.

3.1. *Population, GDP, and GDP growth rate*

This study assumes that population changes by 2050 are exogenous. No population variations are assumed among BAU, APS, and LCET. EAS17 member states except China submitted assumed population changes based on UN projections.

In 2019, the EAS17 population was some 3.89 billion, and is projected to reach approximately 4.37 billion in 2050 at 0.4 percent average annual rate of increase.

Figure 1 shows the general assumptions that Brunei, Cambodia, Lao PDR, and the Philippines will have the fastest average annual population growth rates between 2019 and 2050, ranging from 1.5 percent to

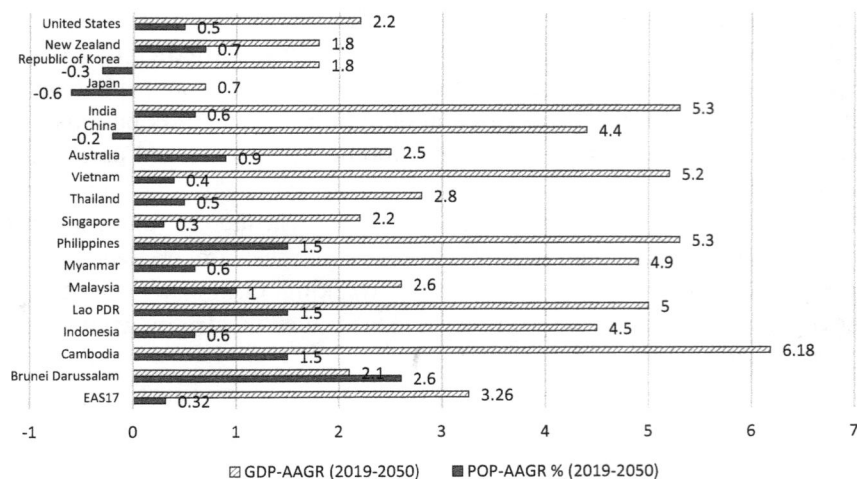

Figure 1.   Average annual EAS17 GDP and population growth rates.

2.6 percent, and that Malaysia, Indonesia, Singapore, Thailand, Vietnam, Australia, India, New Zealand, and the US will see moderate average annual population growth on the order of 0.3–1.0 percent. On the other hand, it is assumed that aging will lead to negative population growth in Japan, China, and South Korea, on the order of −0.2 percent, −0.3 percent, and −0.6 percent, respectively.

Figure 1 also shows assumptions of high long-term economic growth rates in developing EAS17 member states, especially Cambodia, India, the Philippines, Lao PDR, Indonesia, Vietnam, and Myanmar. Brunei, Singapore, Malaysia, and Thailand are expected to have moderate average annual GDP growth rates of 2.1–2.8 percent, as are the developed EAS17 member states, including the US, Japan, South Korea, New

Table 1.   EAS17 GDP and population, 2019–2050.

	GDP (billions USD, 2015)		Population (millions)		Per capita GDP	Per capita GDP
	2019	2050	2019	2050	2019	2050
Brunei Darussalam	14.01	26.55	0.44	0.67	46,700	39,627
Cambodia	20.92	134.14	16.49	26.16	1,269	5,128
Indonesia	1,204	4,710	271	325	4,443	14,492
Lao PDR	18.5	82.8	7.2	11.4	2,569	7,263
Malaysia	364.7	816.6	32.0	43.9	11,397	18,601
Myanmar	74.276	327.01	54.0	65.3	1,375	5,008
Philippines	377	1,847	108.1	171.3	3,488	10,782
Singapore	348.9	683.1	5.8	6.4	60,155	106,734
Thailand	460.8	1,092.5	69.6	82.4	6,621	13,259
Vietnam	162.19	773.93	96.5	108.9	1,681	7,107
Australia	1,346	2,871	25.4	33.0	52,992	87,000
China	14,296	55,025	1,397.7	1,320.0	10,228	41,686
India	2,751	13,447	1,366.4	1,639.2	2,013	8,203
Japan	4,591	5,737	126.3	105.3	36,350	54,482
South Korea	1,638	2,611	51.7	47.3	31,682	55,200
New Zealand	201	348	5.104	6.3	39,381	55,238
United Sates	19,975	38,670	328.2	378.5	60,862	102,166
EAS17	47,843	129,203	3,961	4,370	21,953	37,175

Zealand, and Australia. Rapid growth in China, India, Indonesia, and the US is especially likely to drive energy demand, due to their large economies.

In 2019, total EAS17 GDP was approximately USD 47.9 trillion, accounting for more than half of the global GDP. Given assumptions that regional GDP will grow at an average annual rate of 3.26 percent from 2019 to 2050, Table 1 shows that the total will reach approximately USD 129.2 trillion. China is projected to be the largest regional economy in real GDP by 2050, at approximately USD 55 trillion, followed by the US, at USD 38.6 trillion, India, at USD 13.4 trillion, and Japan, at USD 5.7 trillion.

Average EAS17 real GDP per capita is assumed to increase from USD 21,953 in 2019 to USD 37,175 in 2050. In 2019, per capita GDP ranged from USD 1,269 in Cambodia to more than USD 36,350 in Japan, the US, Singapore, and Australia, and is assumed to range from USD 5,127 in Cambodia to more than USD 102,166 in the US and Singapore in 2050.

3.2. *Thermal electricity generation efficiency*

Thermal electricity generation efficiency, another exogenous assumption of this study, is the amount of fuel required to generate a unit of electricity. The base year 2019 efficiencies were derived from inputs and fuel outputs by fuel type, i.e., coal, gas, and oil, and projected for Brunei Darussalam, Cambodia, Indonesia, Lao PDR, Malaysia, Myanmar, the Philippines, Singapore, Thailand, and Vietnam. Thermal efficiency growth rates were derived from these projections. Assumptions about potential changes in thermal efficiency for the other EAS17 member states were based on the IEEJ Asia/World Energy Outlook 2020, showing increases for new fuels, i.e., hydrogen and ammonia, under LCET, beginning in 2020 and projected toward 2050.

Figures 2–4 show that thermal efficiencies may differ significantly among countries due to differences in technology and fuel input availability and cost, age, and temperature. Thermal efficiencies in ASEAN and the EAS7 are expected to improve considerably over time under BAU as more advanced generation technologies, such as natural gas combined-cycle and supercritical coal-fired power plants, become available. Additional improvements in many countries are also assumed under APS and LCET.

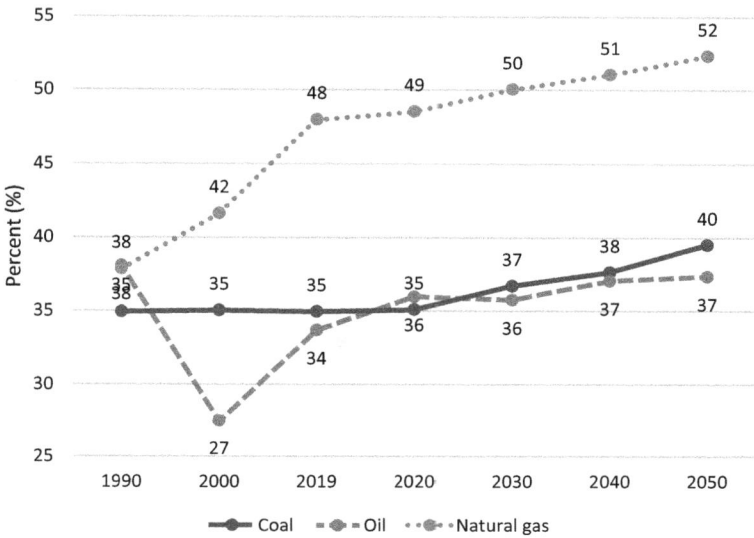

Figure 2. Average EAS17 thermal efficiency under BAU.

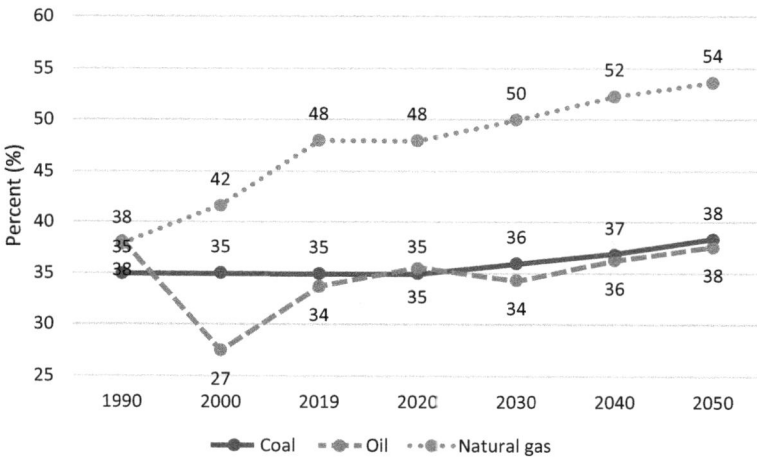

Figure 3. Average EAS17 thermal efficiency under APS.

3.3. *Oil, coal, and natural gas import price assumptions*

Table 2 depicts the oil price assumptions used in the model as adopted by IEEJ from the IEA (2020) world energy model price data. In the Reference

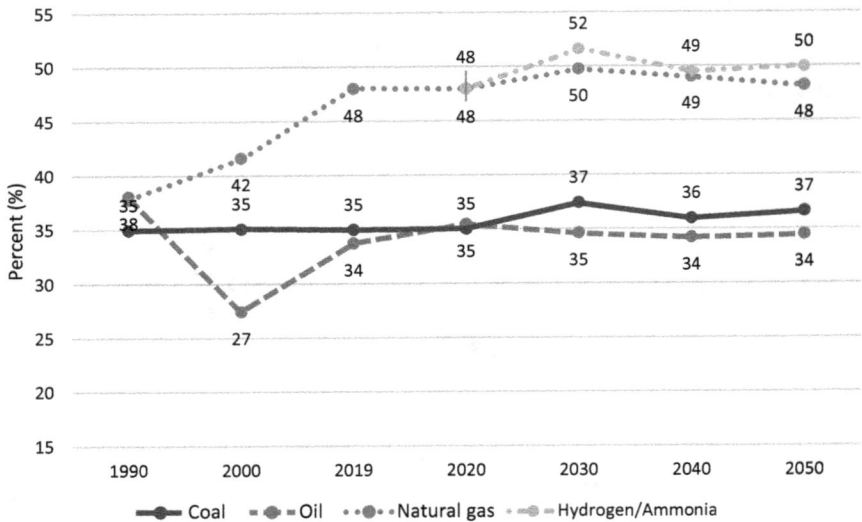

Figure 4.   Average EAS17 thermal efficiency under LCET.

Table 2.   Crude oil, natural gas, and coal import price assumptions (constant prices as of 2020).

Year	Crude oil (USD/bbl)	Coal (USD/ton)	Natural gas (USD/MBTU)		
			US	Europe	Asia
2000	28.66	34.64	4.23	2.71	4.72
2010	79.61	107.14	4.39	6.56	10.91
2015	52.39	79.62	2.60	6.44	10.31
2020	41.00	80.03	2.13	3.25	7.77
2030	80.00	96.00	3.30	7.50	7.60
2040	95.00	97.00	3.80	7.50	7.60
2050	100.00	98.00	3.80	7.40	7.50

Note: While constant price 2020 is used for the energy outlooks of Australia, China, India, Japan, South Korea, and New Zealand, nominal price 2020 is used for ASEAN member states.

Toe: tons of oil equivalent; MBTU: One thousand the British Thermal Unit; bbl: Blue Barrel; Crude oil price assumptions start from 2020.

Source: IEEJ's oil price assumptions (2020).

Scenario, crude oil prices were USD 41/bbl in 2020, USD 80/bbl by 2030, and USD 100/bbl in 2050, rising due to such factors as robust demand growth in non-OECD countries, full economic recovery from the COVID-19 pandemic, emerging geopolitical risks and financial factors, and oil supply constraints reflecting rising oil field depletion rates.

3.4. *Conservation goals and other policy assumptions*

The Working Group was asked to include information on policy assumptions and targets under BAU, APS, and LCET. Some countries in the EAS17 region have clear conservation and emissions reduction targets, which Table 3 summarizes. As the LCET assumption is greater technological efficiency than under APS, Table 3 shows only the APS policy assumptions.

Table 3.   Other assumptions of EAS17 conservation target alignment with NDC under APS.

Country	Assumptions
Australia	Energy efficiency targets a 40 percent improvement between 2015 and 2030. Further 43 percent GHG reduction below 2005 levels by 2050.
Brunei Darussalam	Brunei is committed to a 20 percent GHG reduction from BAU. Brunei also aims to reduce total energy consumption by 63 percent from BAU, and achieve a 10 percent share of renewables in the power mix, by 2035.
Cambodia	Total energy saving of 27 percent from BAU by 2030. Specific fuel efficiency targets by 2050 of 10 percent in coal, oil, gas, biomass, and 20 percent electricity efficiency target. Estimated annual GHG reduction by 2030 under updated NDC scenario by 41.7 percent, approx. 64.6 $MtCO_2$eq.
China	Peak CO_2 emissions before 2030 and carbon neutrality before 2060. President Xi Jinping announced further commitments for 2030 at the Climate Ambition Summit: China will lower CO_2 emissions per unit of GDP by over 65 percent from 2005 levels, increase the share of non-fossil fuels in primary energy consumption to around 25 percent, forest stock volume by 6Gm3 from 2005 levels, and bring total installed wind and solar capacity to over 1.2GkW.

(Continued)

Table 3. (*Continued*)

Country	Assumptions
India	Reduce GDP emissions intensity by 33–35 percent by 2030 from 2005 levels, achieve approx. 40 percent cumulative installed electricity capacity from non-fossil fuel-based energy resources by 2030 with technology transfers and low-cost international finance including Green Climate Fund (GCF), create additional 2.5–3 $GtCO_2e$ carbon sink through additional forest/tree cover by 2030.
Indonesia	Reduce emissions 2020–2030 unconditionally by 29 percent over 2010, 26 percent pledge up to conditional 41 percent against BAU 2030.
Japan	FY2030 aims at a 46 percent GHG reduction from FY2013, and is aligned with the long-term net-zero 2050 goal, striving for 50 percent.
South Korea	Updated GHG reductions target of 24.4 percent from the total national GHG emissions in 2017, which is 709.1 $MtCO_2eq$, by 2030, including increased domestic reduction share through such initiatives as a new coal-fired power plant construction ban.
Lao PDR	Around 60 percent GHG reduction by 2030 over BAU by annual reduced land use, land-use change, and forestry (LULUCF) emissions by 1.1 $MtCO_2eq$, 13 GW hydropower capacity from 5.5 GW, 50,000 energy-efficient stoves, new rapid-transit bus system in Vientiane and new railway to China. Potential 70 percent forest cover, 1 GW wind and solar capacity, 300 MW biomass, energy consumption reduction 10 percent over BAU, conditioned on international support.
Malaysia	Reduce carbon intensity by 45 percent in 2030 from 2005 levels. Around 16 percent of electricity conservation by 2050 in the industry, commercial, and residential sectors. About 16 percent oil conservation by 2050. Replace 5 percent of road transport diesel with biodiesel.
Myanmar	Target saving by 2050 including 20 percent in transport and 10 percent in residential, industry, commercial, and others. Replace 8 percent transport diesel with biodiesel.
New Zealand	Managing the NDC through the emissions budget means net emissions measured across the 2021–2030 target period rather than a single year 2030. Provisional updated NDC budget is 571 $MTCO_2$-e cumulative net emissions between 2021 and 2030 assuming a linear decline from the previous NDC 30 percent GHG reduction target below gross 2005 levels by 2030.
Philippines	Projected 2020–2030, 75 percent GHG emissions reduction, 2.71 percent unconditional, 72.29 percent conditional, in agriculture, waste, industry, transport, and energy over BAU 3340.3 $MtCO_2e$.

Table 3.   (*Continued*)

Country	Assumptions
Thailand	About 30 percent over BAU levels by 2030, up to 40 percent subject to technology development and transfer, financial resources, capacity building support, carbon neutrality by 2050, and net-zero by 2065. Energy efficiency targets by 2050 include transport by 70 percent, residential by 10 percent, commercial by 40 percent, and industry by 20 percent energy demand reduction. Biofuels displace 12.2 percent of transport energy demand.
United States	Around 50–52 percent GHG reductions below 2005 levels in 2030, 100 percent carbon pollution-free electricity by 2035, achievable through multiple cost-effective technology and investment pathways, exploring ways to support decarbonization of international maritime and aviation energy use through domestic action as well as through International Maritime Organization (IMO) and International Civil Aviation Organization (ICAO), invest in new technologies to reduce emissions associated with construction, including for high-performance electrified buildings, support research, development, demonstration, commercialization, and deployment of very low- and zero-carbon industrial processes and products.
Vietnam	Net-zero emissions by 2050, 43.5 percent energy GHG reduction from BAU by 2030, not exceeding 457MCO$_2$e, 91.6 percent from BAU by 2050, not exceeding 101Mt CO$_2$e.

Source: United Nations Climate Change (2022).

4. EAS Energy Outlook

4.1. *Business-as-usual*

4.1.1. *Energy consumption*

Between 2019 and 2050, total EAS17 energy consumption[2] is projected to grow at an average annual rate of 0.9 percent, from 5,318 Mtoe in 2019 to 6,966 Mtoe in 2050, reflecting the aforementioned assumed annual 3.26 percent GDP growth and 0.32 percent population growth. By sector, transport energy demand is projected to grow by 1.3 percent annually, reaching

---

[2]Refers to energy in the form in which it is consumed, including electricity but not fuels and/or energy sources used in electricity generation.

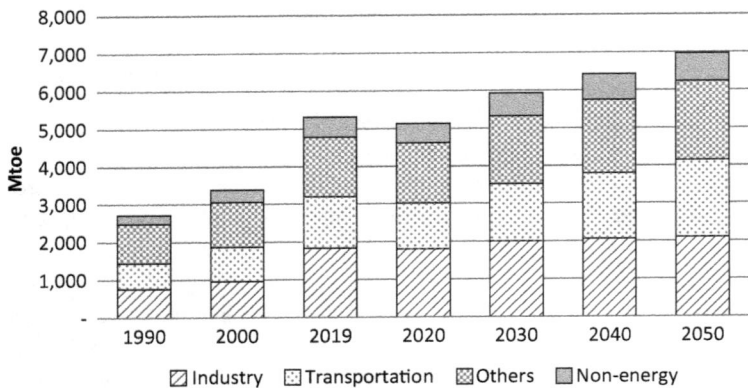

Figure 5.   Energy consumption by sector, BAU, 1990–2050.

Note: BAU: Business-as-usual scenario; Mtoe: million tons of oil equivalent.

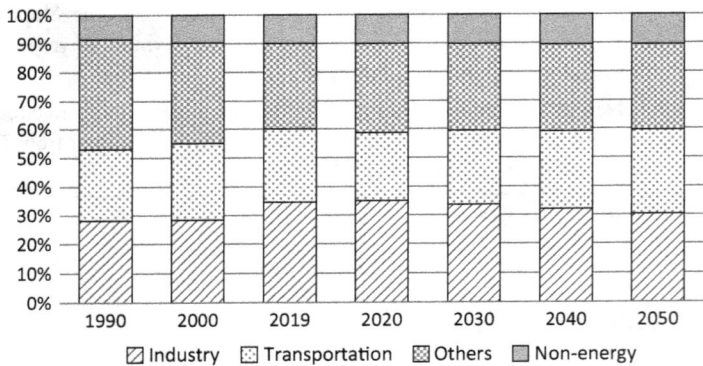

Figure 6.   Energy consumption share by sector, 1990–2050.

29.3 percent projected energy consumption share by 2050. While the annual growth rate for the industry is only 0.4 percent, its energy consumption share is projected to be the largest by 2050, approximately 30.2 percent. Commercial and residential sector, i.e., "Others," demand will exceed industry at 0.9 percent annually, with a 30 percent projected energy consumption share. Figure 5 shows energy consumption by sector, and Figure 6 shares in energy consumption, under BAU in EAS17 from 1990 to 2050.

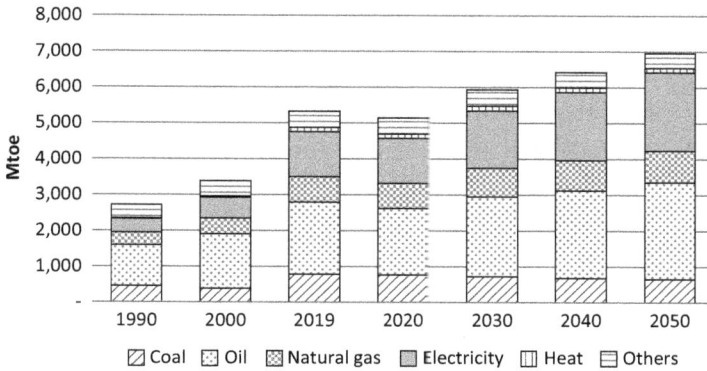

Figure 7.   Energy consumption by fuel, 1990–2050.

Note: Mtoe: million tons of oil equivalent.

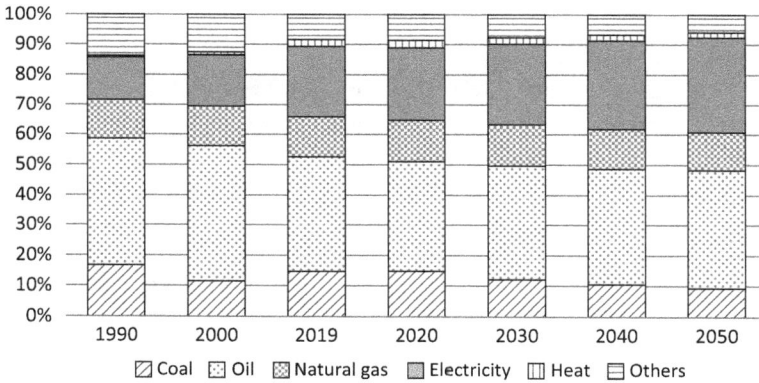

Figure 8.   Energy consumption shares by fuel, 1990–2050.

Figures 7 and 8 show energy consumption and shares by fuel type under BAU from 1990 to 2050. While electricity and natural gas demand in BAU are projected to grow the fastest, by 1.9 and 0.7 percent annually, respectively, their shares will be only 31.6 and 12.5 percent, respectively. Although oil will retain the largest share of total energy consumption at 38.8 percent, it is projected to grow at a lower annual rate of 1.0 percent, reaching 2,706 Mtoe in 2050. Oil share rises slightly from 37.9 percent in 2019 to 38.8 percent in 2050. Coal demand will fall annually at −0.6

percent on average, reaching 653.45 Mtoe in 2050. The share of other fuels such as biomass will decline from 8.4 percent in 2019 to 6.0 percent in 2050. This slow growth is due to the gradual shift from non-commercial biomass to such conventional fuels as liquefied petroleum gas (LPG) and electricity in the residential sector. The share of "heat" demand in energy consumption is also projected to decline from 2.3 percent in 2019 to 1.7 percent in 2050. One reason could be the shift toward more electricity consumption.

4.1.2. *Primary energy supply (BAU)*

Figure 9 shows the primary EAS17 energy supply,[3] which is projected to grow at the same growth rate as energy consumption, i.e., 0.9 percent annually from 2019 to 2050, and to increase from 8,046 Mtoe in 2019 to 10,467 Mtoe in 2050. While coal will still comprise the largest share of the total primary energy supply (TPES) at 32.5 percent in 2050, it is forecast to decline from 39.7 percent in 2019, and it is expected to grow only 0.2 percent annually.

Natural gas is projected to see less moderate growth in 2019–2050, at an annual average rate of 1.3 percent, increasing in share from 16.9 percent in 2019, equivalent to 1,360 Mtoe, to 19.6 percent in 2050,

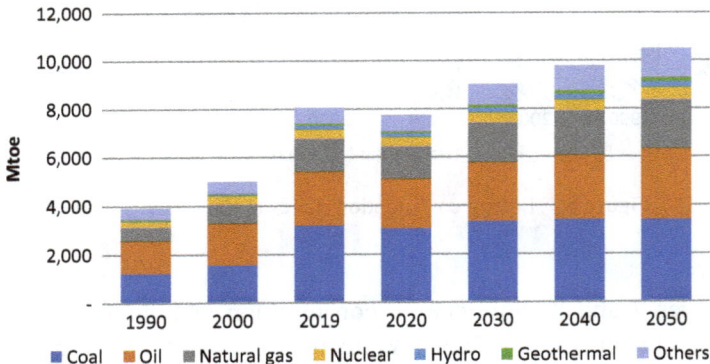

Figure 9.   Primary EAS17 energy supply, 1990–2050.

Note: Mtoe: million tons of oil equivalent.

---

[3]Refers to energy in its raw form, before any transformations, chiefly electricity generation.

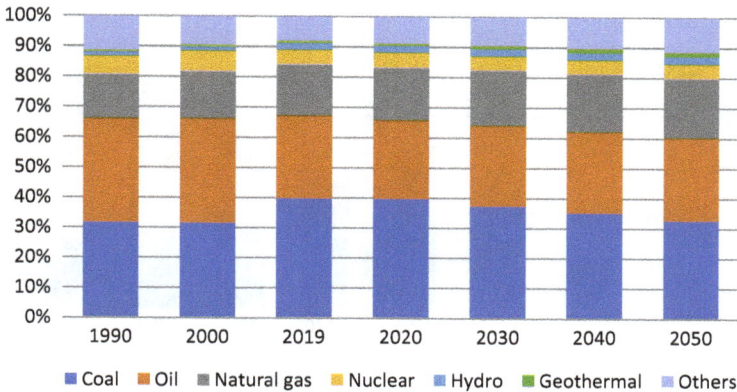

Figure 10.   Primary energy mix shares by source, 1990–2050.

Note: EAS: East Asia Summit.

equivalent to 2,047 Mtoe. Nuclear energy is projected to increase at a slower rate of 0.8 percent annual average, and its share is projected to decline slightly from 4.7 percent in 2019 to 4.6 percent in 2050. It is assumed Japan will resume its nuclear program and that China and India will also resume expanding their capacities. The share of hydropower is projected to rise from 2.2 percent in 2019 to 2.3 percent in 2050. While geothermal is projected to grow fastest, at 3.3 percent annually, its share is projected to remain relatively small, from 0.7 percent in 2019 to 1.6 percent by 2050.

Others, comprising solar, wind, and solid and liquid biofuels, will see annual growth of 1.9 percent from 2019 to 2050, with their share increasing from 8.4 percent in 2019 to 11.6 percent in 2050. Of these, wind and solar will see the largest average annual growth rate, at 4.5 percent, with their combined share increasing from 1.7 percent in 2017 to 5.3 percent in 2050. Figure 10 shows the shares of all energy source in the total primary energy mix over 1990–2050.

4.1.3. *EAS17 energy generation under BAU*

Figure 11 shows EAS17 energy generation, which is projected to grow 1.8 percent annual average, from 2019, equivalent to 16,534 terawatt-hours (TWh), to 2050, equivalent to 28,515 TWh. The growth rate in 1990–2019 was 3.9 percent, more than twice the 2019–2050 projected growth rate.

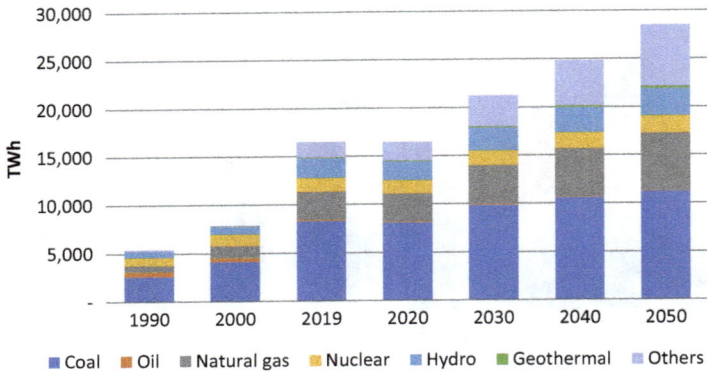

Figure 11.   EAS17 energy generation mix, 1990–2050.

Note: EAS: East Asia Summit; TWh: terawatt-hour.

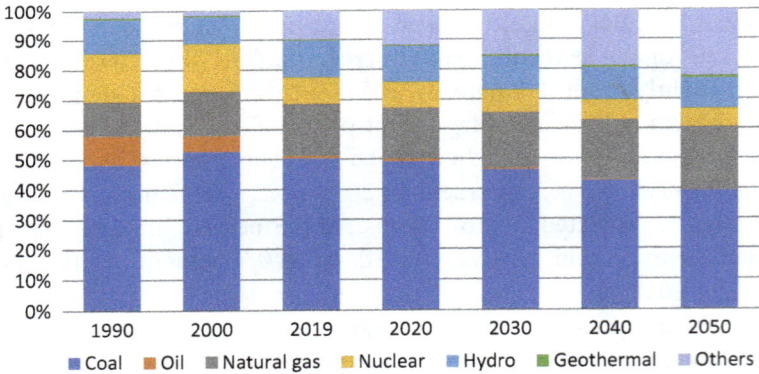

Figure 12.   EAS17 power generation mix shares, 1990–2050.

Figure 12 shows energy source shares in electricity generation from 1990 to 2050. Coal-fired generation is projected to decline from its commanding 50.5 percent share in 2019 to a still majority share of 39.5 percent in 2050. The natural gas share is projected to increase from 17.5 percent in 2019 to 20.8 percent in 2050. Nuclear is forecast to decrease from 8.8 percent in 2019 to 6.5 percent in 2050. Geothermal had a 0.3 percent share in 2019, projected to reach 0.8 percent in 2050. Other sources, i.e., the aforementioned wind, solar, and biomass, will have the highest average annual growth rate, at 4.5 percent. The share of combined wind, solar, and biomass in the power mix is projected at 22.4 percent in 2050, up from 9.9 percent in 2019. Oil will drop from 0.7 percent share in 2019 to 0.1 percent in 2050, with an expected average annual decline of −4.1 percent due to higher costs and increasingly limited use. The share of hydropower is also projected to decrease, from 12.2 percent in 2019 to 9.9 percent in 2050, at 1.1 percent average annual growth.

4.2. *Comparison of BAU, APS, and LCET*

4.2.1. *EAS17 BAU, APS, and LCET energy indicators*

Figures 13 and 14 show end-use and primary energy intensity from 1990 to 2050 for BAU, APS, and LCET. The EAS17 end-use energy intensity, in toe/million 2010 USD, is projected to decline 51.4 percent under BAU, 59.5 percent under APS, and 67.6 percent under LCET in 2050 from 2019 levels. In general, end-use energy intensity has improved, pointing to gradual improvement of energy efficiency in such sectors as industry, transportation, commercial, and residential. The end-use energy intensity

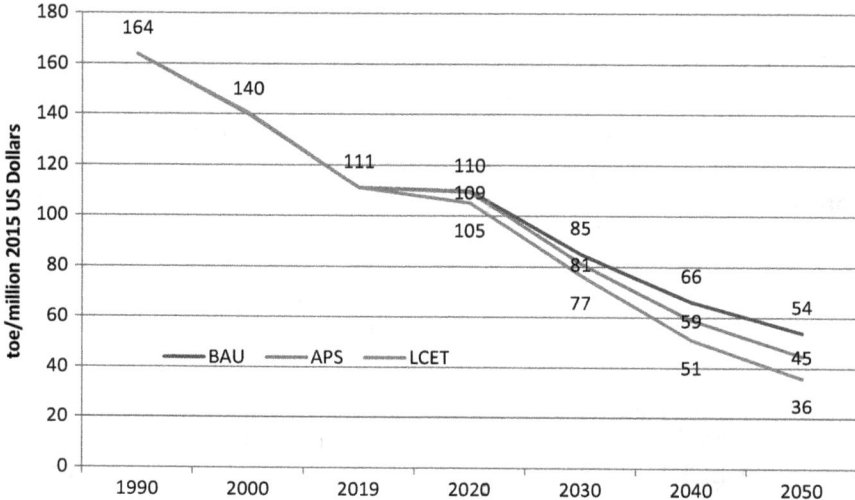

Figure 13.   End-use energy intensity vs. end-use energy consumption per unit GDP (toe/million 2015 USD).

Note: BAU: Business-As-Usual scenario; APS: Alternative Policy Scenario; LCE: Low-Carbon Energy Transition.

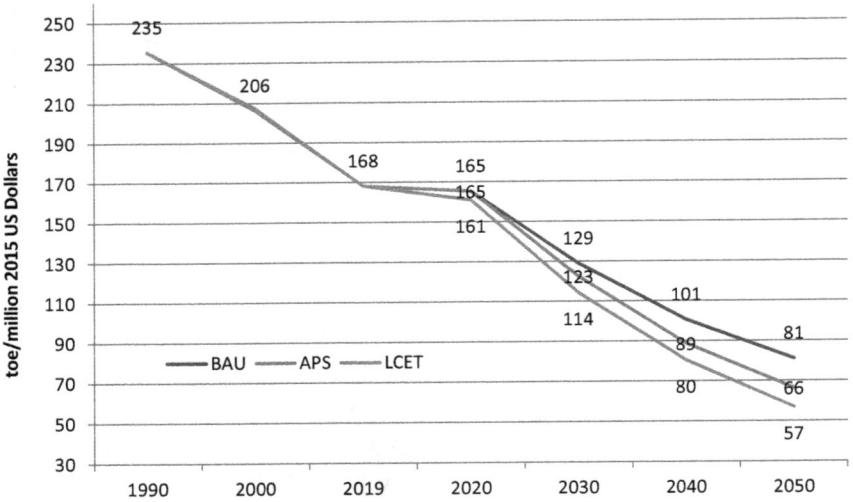

Figure 14.   Primary Energy intensity versus final energy consumption per unit of GDP (toe/million 2015 USD).

Note: BAU: Business-As-Usual scenario; APS: Alternative Policy Scenario; LCET Low-Carbon Energy Transition.

indicator shows how the economy becomes more efficient in end-use energy consumption per unit of GDP, excluding the energy sector.

Primary energy intensity measures whole economy efficiency in turning a unit of energy into a unit of GDP. It also shows energy generation efficiency and end-use energy consumption sectors. Figure 14 shows that the EAS17 is projected to see declines of 51.8 percent under BAU, 60.7 percent under APS, and 66.1 percent under LCET in 2050 from 2019 levels.

Improved carbon intensity, in t-C/million 2015 USD, also shows this improved primary energy intensity.

GDP carbon intensity is also forecast to drop. Figure 15 shows declines of 55.56 percent under BAU, 75.2 percent under APS, and 93.2 percent for LCET, in 2050 from 2019 levels, explained by the shift from fossil fuels to renewables and other clean technologies.

Primary energy supply carbon intensity measures a country or region's performance in terms of CO_2 emissions per unit of energy use, with a shift toward renewables and other clean energy technologies driving it lower. Primary energy intensity is crucial to evaluate gross intensity

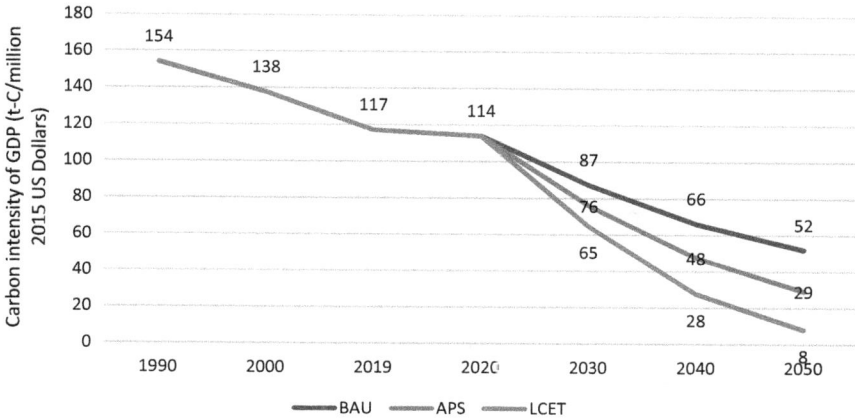

Figure 15.   GDP carbon intensity (t-C/million 2015 USD).

Note: BAU: Business-As-Usual scenario; APS: Alternative Policy Scenario; LCET: Low-Carbon Energy Transition.

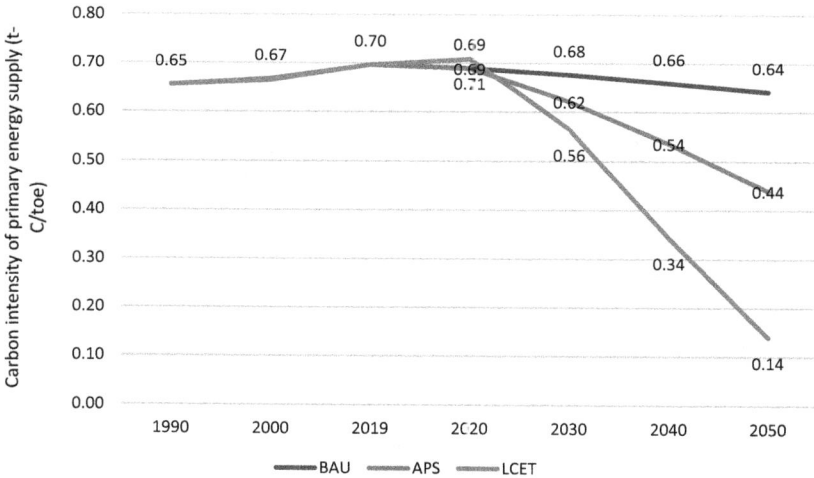

Figure 16.   Primary energy supply carbon intensity (t-C/toe).

Note: BAU: Business-As-Usual scenario; APS: Alternative Policy Scenario; LCE: Low-Carbon Energy Transition.

versus the environment because some ASEAN member states, such as Lao PDR, export coal-fired electricity, giving rise to high primary and low-end use energy intensity. Figure 16 shows that this carbon intensity is projected to decline in EAS17 by 8 percent under BAU, 37 percent under APS, and 78 percent under LCET, in 2050 from 2019 levels. Achieving LCET will accordingly make the region's energy systems that much cleaner.

Energy demand per capita is projected to increase 15 percent under BAU in 2050 from 2017 levels, while decreasing 3.7 percent under APS and 20.8 percent for LCET in 2050 from 2019 levels. These decreases can be attributed either to greater conservation awareness or a response to higher effective energy prices. Projected energy demand per capita is expected to decline as consumers reach energy consumption saturation, given that EAS17 includes the aforementioned developed economies of the US, Japan, South Korea, China, Australia, and New Zealand, with corresponding high per capita energy use. It is conversely expected to rise in developing economies, however, as rising standards of living drive purchases of such energy-consuming devices as vehicles and appliances.

4.2.2. *End-use energy consumption under BAU vs APS and LCET*

Figure 17 shows total end-use energy consumption under APS is projected to rise from 5317.5 Mtoe in 2019 to 5773.8 Mtoe in 2050, which is still 1,192 Mtoe or 17 percent lower than under BAU. The same indicator is likewise predicted under LCET to be 2,321 Mtoe or 33.3 percent lower than under BAU. These results can be attributed to energy efficiency plans for both supply and demand including deploying innovative technologies.

Potential EAS17 total end-use energy consumption savings of 1,192 Mtoe under APS and 2,321 Mtoe under LCET in 2050 are almost three and six times, respectively, ASEAN's 2019 total end-use energy consumption of 447 Mtoe. These figures can largely be attributed to improvements in the transportation, industry, commercial, and residential sectors.

Figure 18 shows end-use energy consumption by sector under BAU, APS, and LCET. Transportation shows the largest reductions, at 24 percent under APS and 46 percent under LCET, followed by Others, at 16 percent under APS and 38 percent under LCET, and industry, at 18 percent under APS and 28 percent under LCET. While non-energy demand will decline by 0.2 percent under APS over BAU, it will increase by a like percentage under LCET, which can be attributed to hydrogen and ammonia fuels. It is anticipated that these will be used in the power co-generation of energy with coal and natural gas to reduce GHGs.

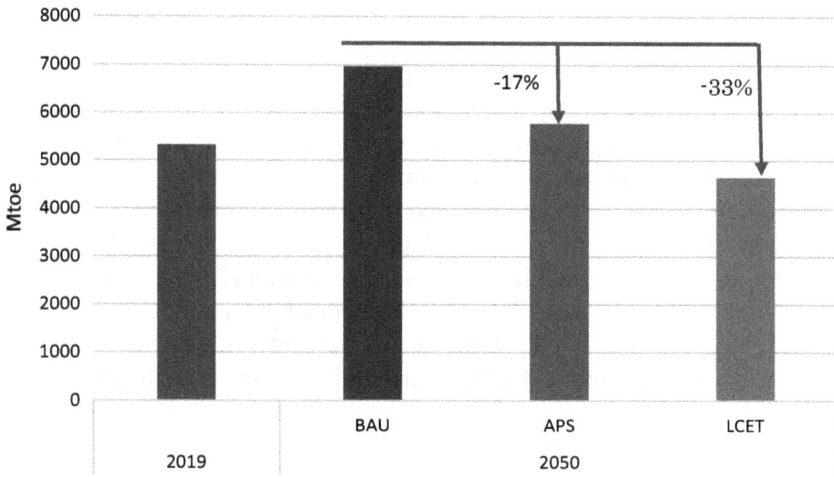

Figure 17. Total end-use energy consumption under BAU vs APS and LCET.

Note: BAU: Business-As-Usual scenario; APS: Alternative policy scenario; LCET: Low-Carbon energy transition.

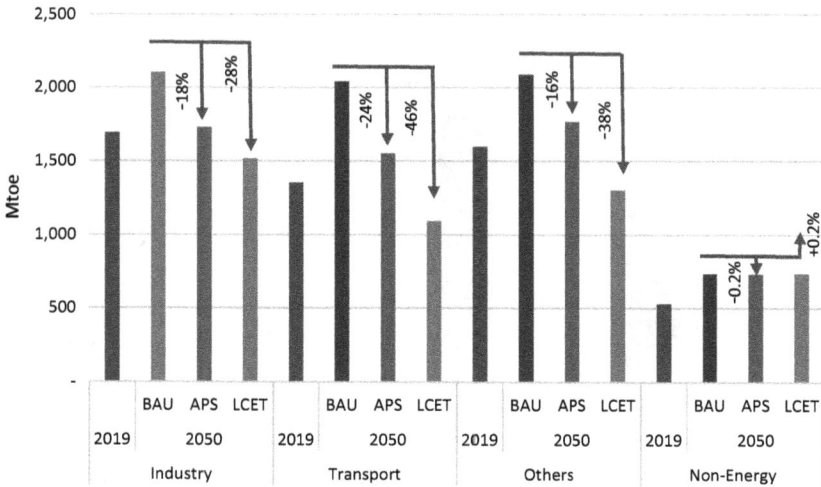

Figure 18. End-use energy consumption by sector, BAU vs APS and LCET.

Note: BAU: Business-As-Usual scenario; APS: Alternative policy scenario; LCET: Low-Carbon Energy Transition.

4.2.3. *Primary energy supply under BAU vs APS and LCET*

Anticipated LCET energy savings accrue to innovations in supply, as well as the aforementioned shift from coal-fired energy to the more efficient gas-fired combined cycle and the previously described combinations of gas with hydrogen and coal with ammonia, as well as the greater shares of renewables and clean technologies, which also reduce GHGs.

Figure 19 shows that the total primary energy supply (TPES) is projected to grow from 8,046 Mtoe in 2019 to 10,467 Mtoe under BAU, 8,558 Mtoe under APS, and 7,347 Mtoe under LCET in 2050. Anticipated total savings are thus 1,909 Mtoe or 18 percent under APS and 3,120 Mtoe or 30 percent under LCET, attributable to improvements in both energy generation and such end-use energy consumption sectors where efficiencies are expected as transportation, industry, residential, and commercial.

Figure 20 shows the primary energy supply by source, indicating that LCET includes such new clean fuels as hydrogen and ammonia in the supply mix, and the aforementioned Others category, comprising solar, wind, and biomass, is projected at 2,271 Mtoe, greater than the 2,047 Mtoe of natural gas supply under BAU. Figure 21 shows primary energy

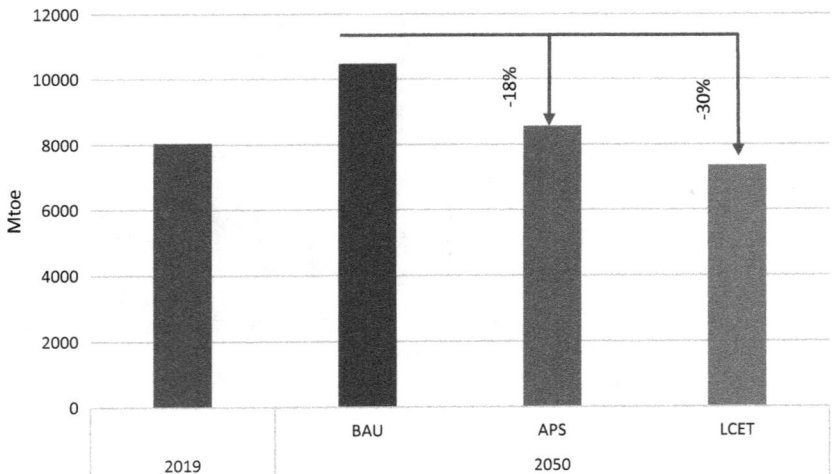

Figure 19.   Total primary energy supply under BAU vs APS and LCET.

Note: BAU: Business-As-Usual scenario; APS: Alternative Policy Scenario; LCET: Low-Carbon Energy Transition.

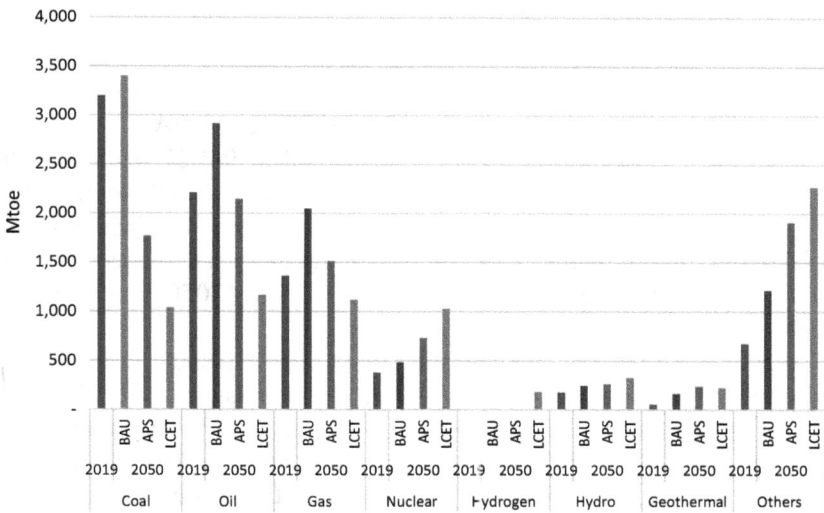

Figure 20.   Primary energy supply by source, BAU vs APS and LCET.

Note: BAU: Business-As-Usual scenario; APS: Alternative Policy Scenario; LCET: Low-Carbon Energy Transition.

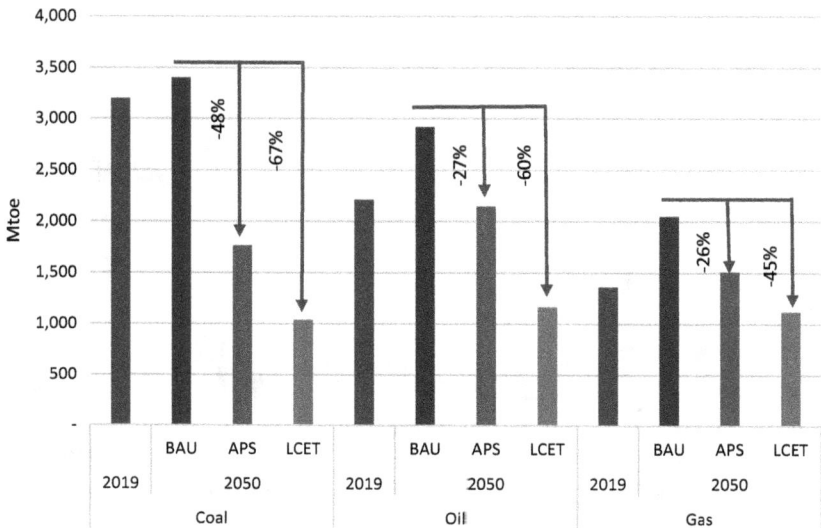

Figure 21.   Primary energy supply by fossil fuel source, BAU vs APS and LCET.

Note: BAU: Business-As-Usual scenario; APS: Alternative Policy Scenario; LCET: Low-Carbon Energy Transition.

supply growth rates, projected as 0.2 percent under APS and −0.3 percent under LCET annual average in the study period, versus 0.9 percent under BAU. Coal demand will fall the most, 1,636 Mtoe or 48 percent under APS and 2,364 Mtoe or 67 percent under LCET versus BAU.

Figure 21 shows that oil TPES is expected to fall 27 percent under APS and 60 percent or 1,749 Mtoe under LCET, attributable to most sectors shifting toward electricity, especially transportation, where EVs are projected to roll out by the early 2040s. Natural gas is also expected to fall 2 percent under APS and 45 percent under LCET by 2050, albeit remaining significant thereafter.

Figure 22 shows how the adoption of nuclear and such renewables as hydropower, geothermal, and solar, wind, and biomass, i.e., the aforementioned Others category, is projected to increase significantly under APS and LCET. Nuclear is projected to increase 51 percent under APS and 112 percent under LCET, showing how vital it is to East Asia's future energy mix and potentially to ASEAN.

While hydropower is fully developed in some countries, others are also forecast to tap its potential, with a predicted 33 percent increase under LCET over BAU. Geothermal is also forecast to increase BAU

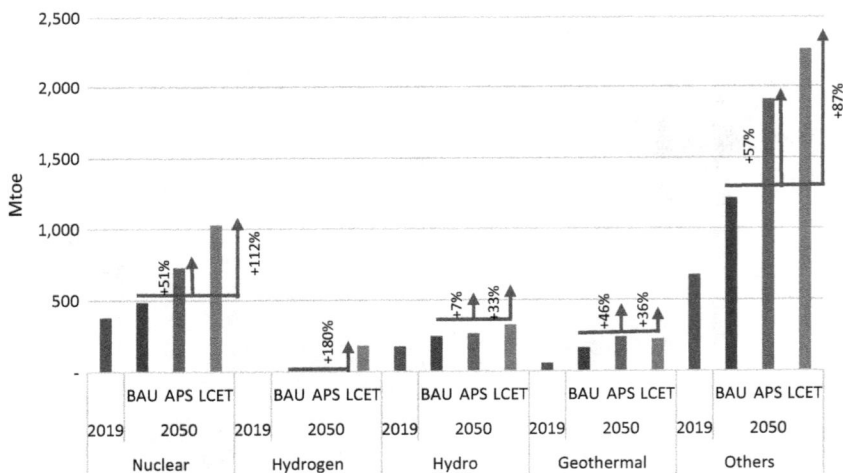

Figure 22. Primary energy supply by clean fuel and renewable sources, BAU vs APS and LCET.

Note: BAU: Business-As-Usual scenario; APS: Alternative Policy Scenario; LCET: Low-Carbon Energy Transition.

cases by 46 percent under APS and 36 percent under LCET, the latter result being due to geothermal fields eventually being exhausted.

It is expected that solar, wind, and biomass, again lumped in the Others category, will increase 57 percent under ASP and 87 percent under LCET. Solar's particular abundance across ASEAN makes it a game-changer in regional energy system decarbonization.

4.3. *Energy generation: BAU vs APS and LCET*

Figures 23 and 24 show that coal and natural gas are predicted to dominate the power generation mix output, contributing 11,268 TWh or 39.5 percent and 5932 TWh or 20.8 percent, respectively, in 2050. The Others category is replaced with individual entries for solar, wind, and biomass under LCET, and its output is predicted to change from 6,380 TWh or 22.4 percent under BAU to 12,543 TWh or 45.6 percent under APS in 2050. Under LCET, the output figures are 6,752 TWh or 21 percent for solar, 8,298 TWh or 25.8 percent for wind, and 2,003 TWh or 6.2 percent for biomass in 2050.

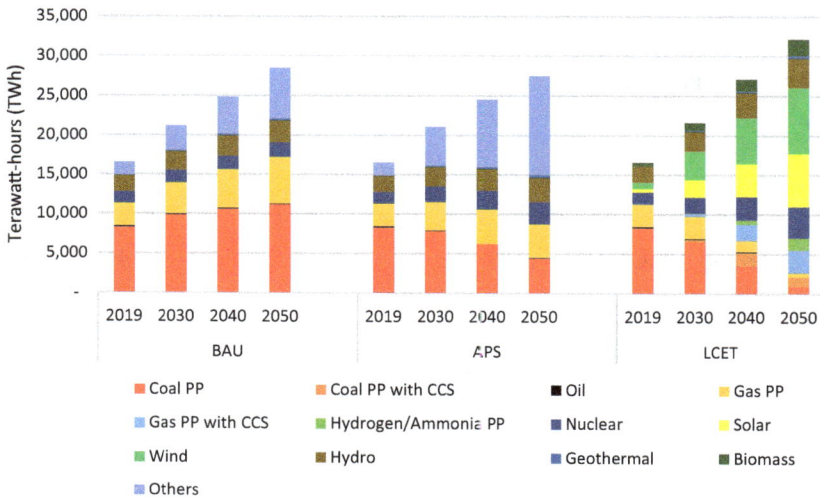

Figure 23.   Energy generation, BAU vs APS and LCET (Terawatt-hour [TWh]).

Note: BAU: Business-As-Usual scenario; APS: Alternative Policy Scenario; LCET: Low-Carbon Energy Transition; PP: Power Plant.

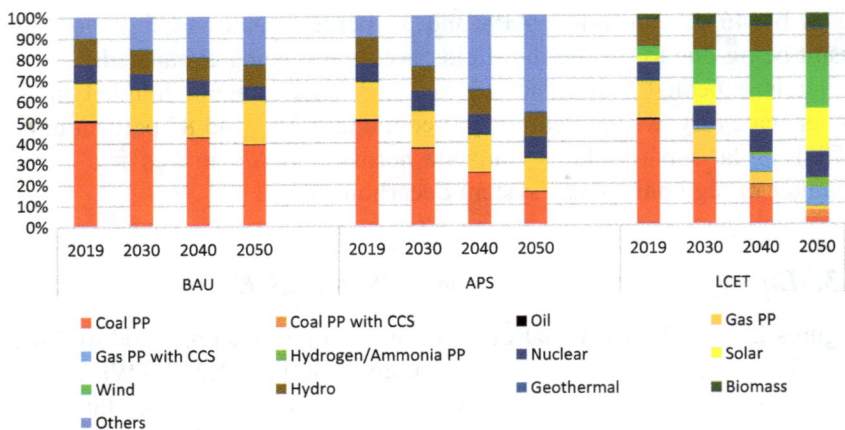

Figure 24.   Energy generation, BAU vs APS and LCET (%).

Note: BAU: Business-As-Usual scenario; APS: Alternative Policy Scenario; LCET: Low-Carbon Energy Transition; PP: Power Plant.

Figure 24 shows hydrogen and ammonia are forecast to enter the generation mix by 2030, increasing in share from 0.1 percent to 2 percent in 2040 and 4.8 percent in 2050. Other clean energy, such as hydropower, geothermal, and nuclear will reach shares under LCET of 11.6 percent, 1 percent, and 12.3 percent, respectively, contributing significantly to decarbonization.

4.4. *CO_2 Emissions from energy consumption*

Fossil fuels emit GHGs into the atmosphere, causing global climate change. We are already seeing the impact, in extreme temperatures, sea level rise, and such natural disasters as severe storms and landslides, resulting in loss of life and property damage. Leaders of 195 countries therefore adopted the Paris Agreement, the first binding global climate treaty, at the Conference of the Parties 21 (COP21) in December 2015. It describes an action plan to put the world on track to limit global temperatures to below 2°C, preferably 1.5°C, above pre-industrial levels by the 2050s.

Figure 25 shows CO_2 emissions under BAU, APS, and LCET, which are projected to reach 6709Mt-C under BAU, 3760Mt-C or 44 percent

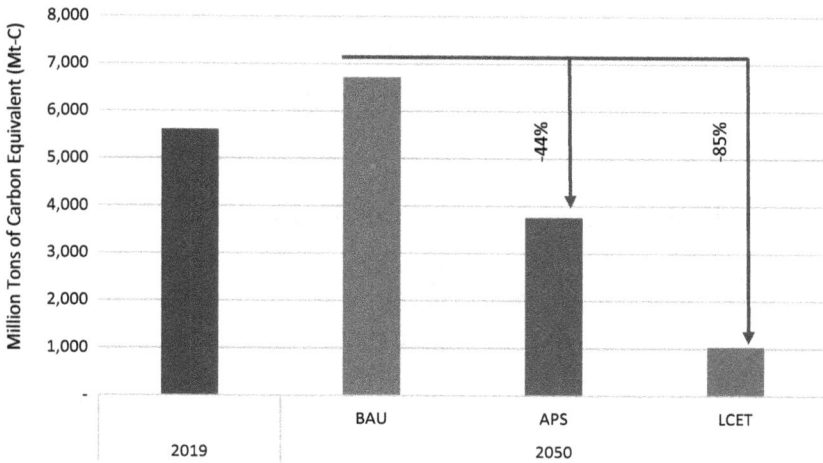

Figure 25.　Total CO_2 emissions, BAU vs. APS and LCET.

Note: BAU: Business-As-Usual scenario; APS: Alternative Policy Scenario; LCET: Low-Carbon Energy Transition.

under APS, and 1,026 Mt-C or 85 percent under LCET in 2050. The latter is chiefly attributable to greater energy efficiency, more renewables, clean fuels, and innovation. Average annual emission growth rates under APS 2019–2050 are −2.1 percent for coal, −0.3 percent for oil, −0.4 percent for natural gas, and −1.3 percent for fossil fuels overall. Under LCET, these are −6.2 percent for coal, −3.9 percent for oil, −5.5 percent for natural gas, and −5.3% for fossil fuels overall. Note also that CO_2 emissions in 2050 are below 2019 levels of 5,606 Mt-C under APS and LCET, in line with Intergovernmental Panel on Climate Change (IPCC) findings that global CO_2 emissions must decline 45 percent from 2010 levels by 2030 to meet the minimal Paris Agreement goal of no more than 2°C above pre-industrial levels and that the aspirational goal of 1.5°C above pre-industrial levels, which the Paris Agreement parties agreed to "pursue efforts" to meet, will require zero-emissions between 2030 and 2050 (IPCC, 2015). It is understood, however, that EAS, and especially ASEAN member states, must balance the priority of abating climate change with that of energy access and affordability. Clean use of fossil fuel through innovations such as clean coal technology (CCT) and the aforementioned co-combustion of coal and natural gas with ammonia and hydrogen, together with CCUS, will play a central role in carbon sinks the world over. Note that this does not include potential natural carbon sinks, which may work to the advantage of some developing ASEAN member states.

5. Conclusions and Recommendations

Although the COVID-19 pandemic significantly affected the global economy as well as EAS17 from 2020 to 2022, it is projected that East Asia and ASEAN will recover beginning in 2023. Given that sustained economic growth in EAS17 is key to the improved per capita incomes and well-being of regional populations numbering in the hundreds of millions, the post-pandemic era forecast is for renewed growth and corresponding increased energy consumption. Population growth and economic growth alike accordingly contribute to the projected TPES in the studied scenarios, from 8,036 Mtoe in 2019 to 10,457 Mtoe under BAU, 8,497 Mtoe under APS, and 4,795 Mtoe under LCET in 2050. Average TPES annual growth rates in the study period are 0.9 percent under BAU, 0.2 percent under APS, and 0.1 percent under LCET. Energy intensity is forecast to decline from 168 toe/USD million in 2017 to 66 toe/USD million or 60 percent under APS and 79 toe/USD million or 53 percent under LCET in 2050. Emission intensity is also forecast to decline, from 0.70 t-C/toe in 2019 to 0.44 t-C/toe or 37 percent under APS and 0.16 t-C/toe or 77 percent under LCET in 2050. The data show that the economy will adopt cleaner, more efficient energy systems, especially under LCET.

Such economic growth will be accompanied by increasing access to, and demand for, electricity and vehicle ownership. Continued reliance on fossil fuels to meet increasing energy demand may lead to increased GHGs and corresponding climate change unless low-emission technologies are adopted. Even supposing adequate domestic fossil fuel resources, oil imports remain probable, with no assurance of security or affordability. The ERIA Comprehensive Asia Development Plan 3.0 (CADP 3.0), a sustainable development policy document, stated that "sustainability is not only a long-term issue but also responds to immediate and urgent problems." This includes rapidly developing such energy infrastructure as transmission and distribution networks and the Trans-ASEAN Gas Pipeline. The plan forecasts that the region will continue to rely on fossil fuels through 2050 under BAU, even given rising crude oil prices. ASEAN and other developing areas must therefore use all available fuels and technologies in pursuing their respective carbon neutrality goals. This study has accordingly explored options under APS and LCET in which key innovations, clean fuels, and renewables will be adopted significantly by 2050.

5.1. *Summary*

Based on projected changes in socioeconomic factors, energy consumption, and CO_2 emissions under BAU, APS, and LCET, the working group reported the following key findings:

(1) Total end-use energy consumption under APS and LCET is forecast to be lower than under BAU due to the expected implementation of energy efficiency plans and effective deployment of innovations on both supply and demand. Total end-use energy consumption in most sectors is also reduced under APS and LCET versus BAU. Transportation shows the largest reductions at 24 percent under APS, 46 percent under LCET, followed by the "Others" sector at 16 percent under APS, 38 percent under LCET, and industry at 18 percent under APS, 28 percent under LCET. Non-energy demand will decline 0.2 percent under APS, and increase 0.2 percent under LCET, versus BAU.

(2) Total energy generation is projected to grow 1.8 percent annually on average from 16,534 TWh, equivalent in 2019, to 28,515 (TWh) in 2050 under BAU. Coal and natural gas are also predicted to dominate the energy generation mix output under BAU, contributing 11,268 TWh or 39.5 percent and 5,932 TWh or 20.8 percent, respectively, in 2050. "Other" generation output is predicted to change from 6,380 TWh or 22.4 percent BAU to 12,543 TWh or 45.6 percent under APS in 2050. Under LCET, solar, wind, and biomass are expected to contribute 6752 TWh or 21 percent, 8298 TWh or 25.8 percent, and 2003 TWh or 6.2 percent, respectively, to the generation mix in 2050. Figure 24 shows that hydrogen and ammonia energy generation is expected to enter the mix by 2030 with a 0.1 percent share, rising to 2 percent in 2040 and 4.8 percent in 2050. Other clean energy also contributes 11.6 percent for hydro, 1 percent for geothermal, and 12.3 percent for nuclear.

(3) Potential EAS17 end-use energy consumption savings in 2050 of 1,192 Mtoe under APS and 2,321 Mtoe under LCET are almost three and six times ASEAN's total end-use energy consumption in 2019 of 447 Mtoe, chiefly in transportation, industry, commercial, and residential sectors.

(4) TPES savings potential in 2050 is projected at 1,909 Mtoe or 18 percent under APS and 3,120 Mtoe or 30 percent under LCET versus BAU. LCET projections include more clean energy, such as shifting from coal-fired energy generation to the more efficient gas-fired combined cycle and co-firing gas with hydrogen or coal with ammonia, in addition to efficiency and other innovations, and extensive use of renewables and clean technologies, all of which contribute to reduced GHGs.

(5) Figure 21 shows that the TPES of oil is expected to decline by 27 percent under APS and 60 percent or 1,749 Mtoe under LCET from BAU, attributable to most sectors shifting toward electricity consumption, especially transportation, where EVs are projected to roll out by the early 2040s. Natural gas is also forecast to decline 26 percent under APS and 45 percent under LCET from BAU by 2050, although it will remain significant thereafter.

(6) End-use energy intensity, which indicates in toe/million 2010 USD how economies use final energy consumption more efficiently per unit of GDP, excluding transportation, is projected to decline 51.4 percent under BAU, 59.5 percent under APS, and 67.6 percent under LCET in 2050 from 2019 levels. This improvement is attributable to steady efficiency gains across sectors, including the aforementioned industry, transportation, commercial, and residential.

(7) CO_2 emissions from energy consumption in 2050 are projected to decline from 6,709 Mt-C under BAU to 3760Mt-C or 44 percent under APS, and 1,026 Mt-C or 85 percent under LCET. The latter may be attributed to the aforementioned greater energy efficiency, as well as greater adoption of renewables, clean fuels, and innovations.

5.2. *Policy implications*

Based on the foregoing, the Working Group identified five major categories of policy implications, enumerated hereinafter. Appropriate policies will vary among countries based on circumstances, policy objectives, and market structures. Note that the Working Group did not unanimously approve all recommendations.

(1) Energy efficiency action plans in end-use consumption sectors. Industry, transport, and commercial sectors will see the largest energy savings. EAS17 member states must implement EEC action plans on an accelerated timetable. By sector:
  • Industry must increase its resource efficiency, and Energy Service Companies (ESCO) have a crucial role to play in doing so. Some ASEAN member states must therefore adopt ESCOs sooner than perhaps intended, by promulgation in national energy policies of compulsory energy management and auditing by businesses consuming large amounts of energy.
  • Transportation contributes significantly to end-use energy consumption growth, as fossil fuel-powered passenger light-duty vehicles, buses, and trucks persist under LCET, despite projections of significant electrification by 2050. This is attributable to oil remaining more affordable than such alternatives as electric or fuel cell-powered vehicles, as well as inadequate EV infrastructure. By contrast, biofuels are projected to see greater use in internal combustion engines and hybrids alike. Greater efforts to expand global EV infrastructure are thus required to meet demand.
  • Energy efficiency can also be considered a resource, as it frees up energy for other economic purposes and supplies to the general population. EAS17 member states are thus encouraged either to review existing regulations or formulate energy efficiency policies that define legal and organizational bases for energy efficiency initiatives and aim to foster reduced energy consumption across all sectors. Said policies will promote energy efficiency as part of national sustainable development policy by (i) applying initiatives and measures for improvement of energy efficiency, especially end-use; (ii) imposing obligatory conservation schemes; (iii) developing energy efficiency services markets to encourage offering services; and (iv) adopting conducive financial mechanisms and schemes. Policies known to serve these purposes include (i) green building codes for energy-efficient buildings; (ii) standards and certifications; (iii) demand management systems that energy managers and ESCOs can manage, such as household and factory energy management systems; and (iv) improving thermal efficiency in

energy generation by replacing existing facilities with newer, more efficient technologies or constructing new facilities from scratch.

(2) Renewable energy policies. There must be greater use of low-carbon fuels, which may be achieved by increased share of renewables, clean fuels such as hydrogen, and nuclear energy in countries' energy mix. Relevant policies and actions include:

- Setting energy mix targets for shares of such renewables as wind, solar, and biomass. Possible incentives may include feed-in-tariffs (FiTs), renewable portfolio standards, and net metering as circumstances including cost-effectiveness dictate. Also necessary to encourage renewables and energy efficiency alike are such aid frameworks as the Clean Development Mechanism (CDM)[4] and Joint Credit Mechanism (JCM)[5].
- Intermittency also poses significant challenges in integrating renewable energy generation with existing power grids, necessitating such bridging storage technologies as on-site hydrogen production, as well as policies and frameworks such as the foregoing to drive down costs.

(3) Smart grids and other aspects of connectivity:

- Once the ASEAN Power Pool is up and running, guided by regional electricity market rules and procedures, it will benefit ASEAN by reducing energy infrastructure development costs, using regional resources more efficiently, assisting utilities in balancing excess supply and demand, improving access to energy services, encouraging renewable capacity development and integration into the regional grid, and attracting additional investment in regional interconnection by providing a price signal that will serve as a key catalyst for returns.
- Policy reforms will be required in the electricity sector, especially deregulation of national and private ownership rules and procedures so as to harmonize with those of the regional power

---

[4]Allows emissions reduction projects in developing countries to earn certified emissions reduction (CER) credits worth 1 tCO_2e apiece.
[5]A project-based bilateral offset crediting mechanism initiated by the Government of Japan to facilitate low-carbon technology adoption.

pool. This includes reducing market barriers by providing a level playing field through unbundling of market segment ownership, allowing non-discriminatory third-party access to transmission and distribution networks, and gradual removal of subsidies in fossil fuel-based power generation, thus fostering fair market competition by renewables and new technologies. Other necessary policies to attract foreign investment into renewables and clean technologies include such incentives as tax holidays, reduced regulatory burdens, rebates through government subsidies, and guarantees to reduce risk.

- ASEAN and East Asia can benefit significantly from smart grid investment as it manages energy generation and demand alike with integrated technology. It involves a complex arrangement of infrastructure whose functions depend on multiple layers. The first layer is the physical components, covering generation, transmission, distribution, and storage. The second layer is the telecommunications services that monitor, protect, and control the grid, including wide area networks, field area networks, home area networks, and local area networks. The third layer is applications facilitated by data management, including data mining as appropriate. The fourth layer consists of software tools that process information collected from the grid to monitor, protect, and control the infrastructure layer, as well as reinforce the grid to allow renewables integration.

(4) Innovation. Environmental technologies will need to be considered in curbing increasing CO_2 emissions:
  - CCUS will be crucial in this regard. Governments must accordingly encourage ongoing CCUS R&D, including its value chain, to ensure its future feasibility.
  - Hydrogen shows promise for commercialization, despite being more expensive than existing fuels, as it can be extracted from fossil fuels, such as low-grade coal and natural gas, as well as through renewable-powered electrolysis. As with CCUS, adoption of hydrogen and other clean fuels depends on ongoing R&D.
  - ASEAN must encourage technological collaboration and adoption, including the hydrogen value chain.

(5) Supply security policies. The following measures are based on the OECD practice of energy security through increasing oil stockpiling requirements (IEA, 2020):

- Promote regional energy connectivity, such as the trans-ASEAN gas pipeline.
- Diversify import sources.
- Reinforce energy infrastructure, including LNG terminal and gasification plant construction.
- ASEAN may need to evaluate public- and/or private-sector strategic reserves in the near future.

(6) Transition Financing. ERIA recently released "The Technology List and Perspectives for Transition Finance in Asia (TLPTFA)," which aims to ease energy transitions with realistic approaches to facilitating developing Asian countries to reach carbon neutrality by 2050 while simultaneously ensuring energy security, affordability, and accessibility, as well as environmental protection with medium-term energy infrastructure financing (ERIA, 2022):

- This list is of particular reference for financial institutions in assessing potential transition technologies submitted by project developers, up until the stakeholders/regulators, e.g., ASEAN and member state governments, are ready with their own technology roadmaps for net-zero investment.
- Although the current draft of the list is not comprehensive, it does cover key potential transition technologies in the energy sector and upstream thereof, which together account for more than 50 percent of Asia's CO_2 emissions.
- The list is intended to provide information on six key elements mentioned elsewhere for each such technology, including cost and reliability, as well as societal benefits and potential GHG reductions.

(7) Asian CCS/CCUS deployment: As mentioned previously, CCS/CCUS is an important pathway to net-zero emissions. The following recommendations are accordingly offered in an attempt to make it more affordable:

- Proposing captured CO_2 capacity targets in line with the Paris Agreement per IE Net Zero Emissions by 2050 Scenario.
- Providing the foundation for widespread commercialization by

2030 through government encouragement of public- and private-sector R&D.

- Encouraging regional collaboration to remove barriers to CCUS value chain development and achieve economies of scale that reduce costs.
- Providing windows of opportunity for financing to developing countries to build practical CCUS demonstration projects in Southeast Asia and elsewhere in the developing world.
- Identifying inexpensive CCUS opportunities, leveraging oil and gas sector experience, expertise, and resources, especially for CO_2 storage opportunities in depleted fields.
- Developing state policy frameworks promoting CCUS development and deployment.
- CCUS capacity building and knowledge sharing to help policymakers comprehend its constituents, what policies it requires, and how best to adapt them to their particular circumstances.

References

Akira, M. and Han, P. (Ed.) (2022). Decarbonization of thermal power generation in ASEAN countries. ERIA Research Project Report 2022. No. 11. https://www.eria.org/uploads/media/Research-Project-Report/RPR-2022-11/Decarbonisation-of-Thermal-Power-Generation-in-ASEAN-Countries.pdf.

Association of Southeast Asian Nations (ASEAN) Secretariat (2007). Cebu declaration on East Asian ENERGY SECURITY 2007. Jakarta: ASEAN Secretariat. http://www.aseansec.org/19319.htm.

Economic Research Institute for ASEAN and East Asia (ERIA) (2015). Scenario analysis of energy security in the East Asia summit region. Jakarta: ERIA. http://www.eria.org/publications/research_project_reports/FY2014/No.35.html, Retrieved June 15, 2020.

Economic Research Institute for ASEAN and East Asia (ERIA) (2022). Special report of COVID-19 impacts on energy demand and energy-saving potential in East Asia, 2021. https://www.eria.org/publications/special-report-of-covid-19-impacts-on-energy-demand-and-energy-saving-potential-in-east-asia-2021/.

ERIA (2022a). Making CCU/CCUS affordable: Enabling CCUS Deployment in G20 and Beyond. Virtual workshop as side event under ETWG G20 2022. https://www.eria.org/events/making-ccs-ccus-affordable-enabling-ccus-deployment-in-g20-and-beyond/.

ERIA (2022b). Technology list and perspectives for transition finance in Asia. https://www.eria.org/publications/technology-list-and-perspectives-for-transition-finance-in-asia/.

ERIA's Press Release (2023). Sustainable energy financing and mobilization of energy investment and advancing CCUS implementations for energy security in ASEAN. https://www.eria.org/news-and-views/sustainable-energy-financing-and-mobilisation-of-energy-investment-and-advancing-ccus-implementation-for-energy-security-in-asean/.

G20 Indonesia (2022). Energy transitions minister's meeting: Bali Compact, 2 September 2022. http://www.g20.utoronto.ca/2022/G20-Bali-COMPACT_FINAL_Cover.pdf.

Han, P. and Endo, S. (2021). Personal communication on oil price assumptions, November 20, 2021.

International Energy Agency (IEA) (2019). World energy model. Paris: IEA. https://www.iea.org/reports/world-energy-model/macro-drivers.

IEA (2020a). Oil stocks of IEA countries. September, 15, 2020. https://www.iea.org/articles/oilstocks-of-iea-countries.

IEA (2020b). *World Energy Balances*. Paris: OECD Publishing. https://www.iea.org/reports/world-energy-balances-overview#non-oecd-asia.

Kimura, S. *et al.* (Ed.) (2022). Decarbonization of ASEAN energy system: Optimum technology selection model analysis up to 2060. ERIA Research Report FY 2022. No. 05. https://www.eria.org/research/decarbonisation-of-asean-energy-systems-optimum-technology-selection-model-analysis-up-to-2060/.

Kimura, S., Han, P., and Alloysius, J. P. (Ed.) (2023). Energy outlook and energy-saving potential in East Asia. https://www.eria.org/research/energy-outlook-and-energy-saving-potential-in-east-asia-2023/.

OILPRICE.COM (2022). Oil price charts. https://oilprice.com/oil-price-charts/46.

The Institute of Energy Economics, Japan (IEEJ) (2014). Asia/world energy outlook 2014. Tokyo: IEEJ. Accessible at https://eneken.ieej.or.jp/data/5875.pdf.

The Institute of Energy Economics, Japan (IEEJ) (2017). Asia/world energy outlook 2017. Tokyo: https://eneken.ieej.or.jp/data/7303.pdf.

United Nations Climate Change (2022). 2022 NDC Synthesis Report. Accessible at https://unfccc.int/ndc-synthesis-report-2022.

United Nations Climate Action (2023). Cop26: Together for Our Planet. Accessible at https://www.un.org/en/climatechange/cop26.

United Nations Framework Convention on Climate Change (UNFCCC) (2022). Intended Nationally Determined Contributions: Submissions. New York: UNFCCC. Accessible at https://www4.unfccc.int/sites/submissions/indc/Submission%20Pages/submissions.aspx.

Chapter 2

Deployment of CCUS for Future ASEAN Decarbonization: An Energy Justice Perspective

Citra Endah Nur Setyawati

Energy Affairs ERIA, Jakarta Pusat, Indonesia

citra.endah@eria.org

Abstract

Carbon capture, utilization, and storage (CCUS) encompasses a technology with the potential to fill many roles in the pursuit of global energy and climate objectives. There are now some 10 ongoing carbon capture projects in ASEAN countries, with the majority of these initiatives being announced since January 2020. Nevertheless, the implementation of CCUS as a strategy to address climate change has encountered opposition from certain individuals who argue that it perpetuates continued reliance on fossil fuels rather than promoting their substitution with renewable energy alternatives, such as solar and wind power. This study employs energy justice as an analytical tool to analyze social and economic aspects within energy systems to ensure fairness in allocating both benefits and responsibilities associated with energy provision. The results show that potential injustices might occur in developing CCUS technology. Hence, the eight principles of energy justice — availability, affordability, due process, good governance, sustainability, intergenerational equity, intragenerational equity, and responsibility — must be taken into consideration so as to promote inclusiveness and enable broad participation in ASEAN's CCUS development plans.

Keywords: CCUS, energy justice, ASEAN, decarbonization, CO_2 emissions.

1. Background

Urgent and comprehensive measures to mitigate greenhouse gas emissions originating from energy-intensive sectors are needed in order to achieve the global average temperature target of 1.5°C Celsius. Countries have a responsibility to support and stimulate innovation as well as research and development to accelerate the adoption of low- and zero-carbon technologies. One such technology is carbon capture utilization and storage (CCUS), which removes carbon dioxide (CO_2) from sources such as coal-fired power plants with the intention of reuse and restoration of the captured emissions. According to the Intergovernmental Panel on Climate Change (IPCC, 2005), CCUS is used for industrial procedures that separate CO_2 near emission sources.

ASEAN CO_2 emissions accounted for 1610 $MtCO_2$, which are substantially incurred from coal and oil utilization (IEA, 2019). CCUS can play diverse roles in meeting global energy and climate goals (IEA, 2022). More than 10 carbon capture projects are currently under development in ASEAN, the majority of which were announced in January 2020. These projects, located in Indonesia, Malaysia, and Thailand, have a potential total capture capacity of approximately 15 $MtCO_2$ annually by 2030 (IEA, 2022). CCUS can be used to reduce CO_2 emissions and thus reduce the impact of climate change. If CCUS technologies are not used, the United Nations Intergovernmental Panel on Climate Change (IPCC) forecasts that the cost of mitigation would increase by 138% in 2100. CCUS is mentioned as one of three obligatory emission reduction technologies in the four main emission reduction technological pathways in the IPCC's 2018 1.5°C special report.

Nevertheless, the use of CCUS to combat climate change has been met with criticism from some who believe it encourages the ongoing use of fossil fuels rather than their replacement with renewable energy sources, such as solar and wind. While scholars agree that CCUS technology should be included among the options to decarbonize and mitigate climate change, as stated in the latest IPCC report, they claim that renewable energy sources like solar and wind, as well as power storage, are progressing more rapidly than CCUS (Costley, 2022). Critics of carbon capture and storage say it has not been demonstrated to work and that other methods, like the aforementioned solar and wind power, have been more successful at reducing greenhouse gas emissions (Costley, 2022).

As a result, transitions to low-carbon energy systems must take energy justice principles into consideration to ensure that policies and plans offer fair and equitable access to resources and technology. Energy justice is a cross-disciplinary social science research theory constructed from environmental justice theory that aims to apply procedural, distributive, and recognition justice concepts to energy systems and transitions (Wood, 2023a).

The energy justice framework is employed in this study as an analytical tool to understand how values are incorporated into energy systems and ensure that well-informed energy choices are made when selecting CCUS as the technology option for achieving energy transition. From the standpoint of energy justice, this research aims to address the following questions: To what extent do the current CCUS projects consider the key principles of energy justice? And how can ASEAN ensure that CCUS will be affordable, with benefits that can be distributed to future generations in a just manner?

The objectives of this chapter are to understand different and conflicting visions of future adoption of CCUS technologies in ASEAN and examine how energy justice can serve as an analytical tool in CCUS deployment. The analysis in this chapter is based on peer-reviewed journals, government websites, gray literature, and reports from international organizations. Triangulation has been employed when it has been determined that differing sources may have varying degrees of credibility (Stake, 1995).

The remainder of this chapter is organized as follows. Section 2 describes the methodology used, Section 3, results and discussion, examines current CCUS projects to provide a basic understanding of development trends in ASEAN countries, and explains how energy justice can serve as an analytical tool in the case of CCUS deployment, and Section 4 provides the conclusion.

2. Methodology

The theory of energy justice connects energy systems to questions of social justice. It addresses and prevents injustices concerning energy

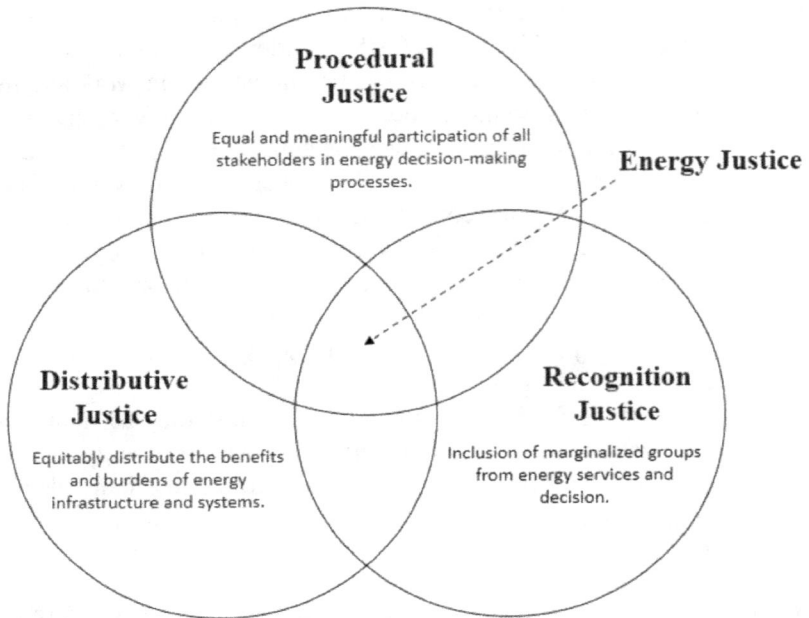

Figure 1.   Core energy justice principles.

Source: Wallsgrove *et al.* (2021), modified by the author.

and society. The framework of "energy justice as a decision-making tool" as used in this analysis aims to facilitate better-informed choices by governments and planners, while the term "energy justice" refers to a global energy system that distributes the benefits and costs of energy services equitably and has transparent, non-discriminatory decision-making (Sovacool and Dworkin, 2015). Figure 1 denotes core energy justice principles, namely procedural, distributive, and recognition justice. Jenkins *et al.* (2016) used the evaluative and normative evaluation questions of energy justice to obtain an understanding of an opportunity to examine the occurrence of injustices, develop novel avoidance strategies, and recognize overlooked segments of society.

Table 1 depicts questions that can be asked to understand the evaluative and normative contributions of the energy justice tenets. To address instances of injustice, it is necessary to (a) ascertain the specific issue at hand (distributive), (b) determine the individuals or groups that are impacted by the injustices (recognition), and (c) devise appropriate tactics and procedures to rectify the injustices (procedural) (Jenkins *et al.*, 2016).

Table 1. Evaluative and normative contributions of energy justice.

Tenets	Evaluative	Normative
Distributive	Where are the injustices?	How should we solve these?
Recognition	Who is ignored?	How should we recognize those who are ignored?
Procedural	Are there fair processes?	Which new processes should be instituted?

Source: Jenkins *et al.* (2016).

2.1. *Procedural justice*

Procedural justice emphasizes "equitable procedures that engage all stakeholders in a non-discriminatory way" (McCauley *et al.*, 2013, p. 2). Put another way, its objective is to ensure equal and meaningful participation of all stakeholders in energy decision-making (Jenkins *et al.*, 2016). It comprises a review of whether and to what degree energy decision-making processes are democratic and inclusive. Procedural justice offers opportunities for seeking solutions in energy technology development by employing appropriate methodologies to actively include stakeholders in decision-making processes, such as disclosures of relevant information and establishment of protocols and mechanisms to ensure participation, openness, and fair treatment (Setyowati, 2021).

2.2. *Distributive justice*

Distributive justice acknowledges the presence of disparities in allocating environmental benefits and burdens, as well as the unequal distribution of corresponding duties. It assesses where "questions about the desirability of technologies in principle become entangled with issues that relate to specific localities" (Owens and Driffill, 2008, p. 4414). Put another way, distributive justice entails equitable allocation of the advantages and costs with the production and consumption of energy. The concept being discussed is inherently spatial in nature, as it pertains to the examination of uneven patterns in relation to physical placement and dispersion of energy resources and negative consequences thereof (McCauley *et al.*, 2013; Setyowati, 2021). Distributive justice concerns access to energy services as well as infrastructure siting (Jenkins *et al.*, 2016). Numerous studies have emphasized that changes inherently result in both beneficiaries and those who are adversely affected, underscoring how critical are the inclusiveness and distributional dimensions of emerging technologies in energy transitions.

2.3. *Recognition justice*

Recognition justice involves recognizing the unique identities and histories of individuals in connection with energy systems and working toward eliminating all types of sociocultural dominance that some groups have over others (Jenkins *et al.*, 2016; McCauley, 2013). When trying to pinpoint sources of injustices, policymakers may fail to consider the full extent of their effects on marginalized groups (McCauley *et al.*, 2013). Fraser (1999) classifies the three basic types of misrecognitions as cultural dominance, non-recognition, and contempt. The fundamental concept underlying recognition pertains to the sufficient acknowledgment of a person or a collective entity. Adequate acknowledgment entails demonstrating respect and recognition toward individuals based on their identities and circumstances. According to Preston and Carr (2018), recognition further entails the equitable representation and thoughtful consideration of the cultures, values, and circumstances of all parties involved. In the case of individuals being excluded from decision-making processes, it is imperative to acknowledge stakeholder groups that experience either favorable or adverse consequences as a result of energy initiatives.

Sovacool and Dworkin (2015) summarized the guiding principles that the government may utilize to inform policy decisions: availability, affordability, due process, transparency and accountability, sustainability, intra- and intergenerational equity, and responsibility (Table 2). The understanding and the adaptation of these aspects in this analysis are presented in Table 3.

3. Results and Discussion

3.1. *Overview of current ASEAN CCUS development*

Interest in CCUS is rising notably in Southeast Asia as well as globally. Its resurgence has been fueled by increased climate pledges from governments and businesses, including aggressive net-zero targets. The investment climate for CCUS has also improved: governments and businesses throughout the world have pledged at least USD12 billion in funding,

Table 2.   Energy justice decision-making tool.

Principle	Description
Availability	People deserve adequate high-quality energy resources.
Affordability	All people, including the poor, should pay no more than 10% of their income for energy services.
Due process	Countries should respect due process and human rights in their production and use of energy.
Good governance	All people should have access to high-quality information about energy and the environment, as well as fair, transparent, and accountable forms of energy decision-making.
Sustainability	Energy resources should not be depleted too quickly.
Intragenerational equity	All people have a right to fair access to energy services.
Intergenerational equity	Future generations have a right to enjoy a good life undisturbed by the damage our energy systems inflict on the world today.
Responsibility	All nations have a responsibility to protect the natural environment and minimize energy-related environmental threats.

Source: Sovacool and Dworkin (2015).

Table 3.   Adaptation of energy justice as a decision-making tool framework proposed by Sovacool and Dworkin (2015), modified by the author.

1. Availability	• Robust and **diversified energy value chain**, resilient in the face of disruptions. • **Technological solutions** that regions utilize to produce, transport, conserve, store, or distribute energy. • Amount of **investment** needed to keep systems functioning.
2. Affordability	• Affordability of energy services in terms of **price and price stability**. • Energy bills that do not overly **burden consumers**. • Available energy is meaningless when unaffordable.
3. Due processes	• Engaging the public in decision-making. • **Involvement of stakeholders** such as businesses in energy governance. • Ensuring **fair and informed consent** of affected communities.

(*Continued*)

Table 3.   *(Continued)*

4. Good governance	• Information, accountability, and **transparency**. • **Reducing corruption**. • **Access** to information on governance.
5. Sustainability	• Meeting present energy needs **without compromising the future**. • **Sustainable use of natural resources** without environmental damage.
6. Intragenerational equity	• **Distributive justice**: Who benefits and who loses? • Individuals have the **right to a minimal set of energy services** for basic well-being.
7. Intergenerational equity	• Distributive justice between **present and future generations**. • Ensuring energy systems are **not damaging to future environments**.
8. Responsibility	• Avoiding social and environmental **negative externalities**. • Responsibility of the Global North.

particularly for CCUS projects since early 2020. Such projects are eligible for an additional USD20 billion in clean energy funding from programs that were initiated in early 2020 (IEA, 2021a).

The IEA (2021a) conducted a comprehensive analysis of emissions originating from power and industrial facilities in Southeast Asia. The primary objective of this study was to ascertain significant hubs of industrial activity that exhibit substantial emissions. The present comprehensive geographical investigation reveals notable emission clusters on the island of Java in Indonesia, with several industrial hubs along the coastal regions of Vietnam and Luzon, which is the largest island in the Philippines.

Figure 2 depicts CO_2 sources in Southeast Asia, where at least seven potential projects have been identified and are in early development stages in Indonesia, Malaysia, Singapore, and Timor-Leste. The CCUS Gundih Project in East Java, Indonesia, is regarded as the first CCUS pilot project in ASEAN. Its objective is to store CO_2 from the Gundih Gas Field, which contains approximately 21% CO_2 that is flared from the Gundih Centre Processing Plant (CPP). Its total CO_2 storage is on the order of 3Mt for 10 years (IEA, 2021a).

Figure 3 projects regional CCUS deployment based on the IEA Sustainable Development Scenario, which is consistent with achieving

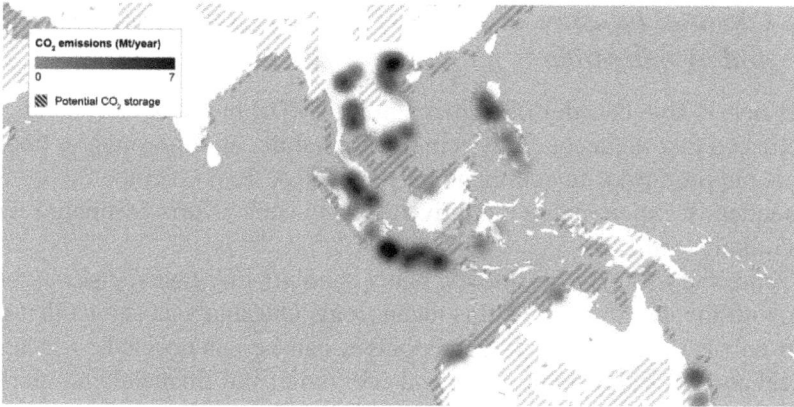

Figure 2. CO_2 sources in Southeast Asia.
Source: IEA (2021a).

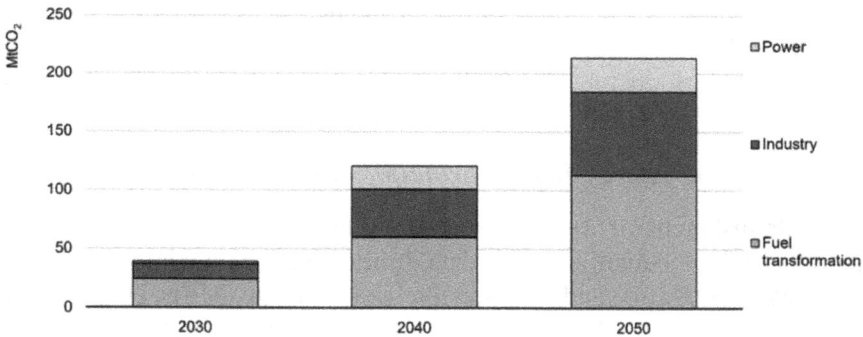

Figure 3. CO_2 sources in Southeast Asia under sustainable development scenarios.
Note: Values shown are from the IEA Sustainable Development Scenario; corresponding CCUS deployment levels are generally higher in the Net Zero 2050 Roadmap.
Source: IEA (2021a).

net-zero emissions globally by 2070. This would be the minimal CCUS deployment required in Southeast Asia to meet the more aggressive IEA (2021b) Net Zero by 2050 Scenario. The IEA (2021a) report also showed that Southeast Asia's CO_2 storage capacity is 390Gt, with 54% offshore and 46% onshore estimated to be available in the region.

3.2. *Potential injustice in CCUS technology development and its mitigation*

After nearly five decades of implementing CCUS in many nations, it can be inferred that the technology remains immature, and thorough deliberation is required prior to commencing projects. ASEAN member states are consequently experiencing a relative delay in such efforts compared with the Global North.

From the CCUS case studies, many scholars mentioned risks associated with the deployment of this technology. Our analysis shows that the main barriers that hinder its development can be categorized as technological and environmental, social and economic, and financial and market risks. The study by Wang *et al.* (2021) revealed that greater plant capacities make CCUS projects more likely to be canceled or postponed, with a capacity increase of 1 $MtCO_2/y$ raising the probability of failure by approximately 50%. Gradual upscaling and increased policy support, particularly for demonstrations of CCUS viability, while also building a market through carbon pricing, would help address the current imbalance between risk and return. Table 4 illustrates the potential for injustice arising from the introduction of CCUS.

3.3. *Assessing energy justice principles in CCUS deployment*

The establishment of a robust justice system is crucial in effectively addressing the inequalities and vulnerabilities connected with the responsibilities of mitigating and adapting to climate change (Moss, 2009). Such a system should prioritize protecting human rights and be grounded in a theory of social justice that encompasses collective solutions. The organization of this section is based on the research questions and the applicable framework, with a focus on the components to be analyzed. It is imperative that all aspects of the ASEAN case study be comprehensively analyzed, given that CCUS can be regarded as a nascent technology within the region.

This chapter builds on the three tenets cited earlier by providing a conceptual perspective and a list of energy justice principles. Using such frameworks to evaluate CCUS technology development can aid in uncovering its implications for justice. The analysis of this study is conducted by integrating energy justice with current CCUS development initiatives,

Table 4. Potential injustices of CCUS and mitigations thereof.

Procedural Justice		
Possible injustice to be avoided	**Description**	**Mitigations**
Unjust decision-making processes	Lack of procedural fairness and trust in decision-making resulting in low acceptance levels by local communities in CCUS projects (ter Mors and van Leeuwen, 2023).	Giving local communities a voice can boost the sense of fairness in decision-making procedures, foster trust in the project developer, and result in greater CCUS project acceptability (ter Mors and van Leeuwen, 2023).
Vested interests, motivated by economic and political goals rather than justice	Political support can be influenced by vested interests (Storrs *et al.*, 2023).	A clear "political agenda" has helped CCUS gain traction in Canada, Australia, the US, Norway, the UK, and the Netherlands (Storrs *et al.*, 2023; Bui *et al.*, 2018).
Limited access to public information and consultation	In Australia, local communities lacked access to accurate project information and the means to bargain and interact with project partners (Nielsen *et al.*, 2022).	Right to information and capacity building of local communities (Nielsen *et al.*, 2022).
Lack of evidence-based policymaking	No clear vision for CCUS deployment in ASEAN, making transition from evidence to policies difficult. Only Singapore envisions establishing a carbon capture hub, targeting 2MTPA of CO_2 capture capacity by 2030 (IEEFA, 2022).	Create "strategic vision" for CCUS technologies to clarify the rules and give certainty to investors (Taylor, 2022). Clear CCUS roadmap involving both government and industry will be needed to deploy large-scale CCUS projects (IEA CCUS Investor Roadmap, 2022).

(*Continued*)

Table 4. (*Continued*)

Distributive Justice		
Injustice	**Description**	**Potential mitigation actions**
High long-term costs to consumers and industries	Substantial energy penalty and CCUS retrofit of coal-fired power plants could decrease output by 2–30% (Wei *et al.*, 2021), resulting in high-cost low-carbon electricity.	Incentives and subsidies sufficient to offset additional costs for fossil fuel-based industries are needed if large-scale CCUS is to be implemented (Wei *et al.*, 2021). High carbon prices should be adopted to promote CCUS growth (Ghilloti, 2022).
Safety risks	Safety risks in capture, i.e., plant facilities and equipment failure; transportation, i.e., CO_2 leakage from pipelines, ships, and tanks; utilization, i.e., well collapses, pipe wall corrosion, aging equipment; and storage, i.e., leakage from natural traps and injection, and CO_2 storage security and duration in geological formations (Chen *et al.*, 2022).	Technical management standards and international unified regulatory framework (Chen *et al.*, 2022).
Health risks	Threats to respiratory health from pollution; nitrosamines and nitramines causing cancer; domino effects from other harmful substances; severe casualties and property losses; hazards to offshore platform workers (Chen *et al.*, 2022).	Establish health risk assessments and assign and regulate responsibility (Dillman and Heinonen, 2022).
Environmental risks	Emission of harmful substances, i.e., NOx and NH_3; liquid and solid waste; freshwater toxicity; air pollution; damage to soil, vegetation, and	Study environmental risk assessments and lessons learned from oil and gas projects, geothermal projects, and natural disasters (Chen *et al.*, 2022).

Injustice	Description	Potential mitigations
	natural habitats; oxygen depletion; groundwater hazards (Chen et al., 2022).	Environmental management standards and international unified regulatory framework (Chen et al., 2022).
Unrealistic notion of financial security	CCUS scale and cost intensity make widespread adoption unlikely; also, insufficient price attached to carbon emissions (IEEFA, 2022a).	Concessionary financings, bilateral initiatives, substantial public funding, and supportive regulations (IEEFA, 2022a).
Stranded assets; reliance on gas and coal infrastructures	In coming decades, Southeast Asian market will likely be limited around gas processing (IEEFA, 2022a); 7 out of 13 CCUS projects underperformed, two failed, and one was mothballed (IEEFA, 2022b).	Assess likely ASEAN CCUS scenarios.

Recognition Justice

Injustice	Description	Potential mitigations
Climate change discrimination — displacement of climate costs to future generations	Generations to come will face proportionally greater chances of climate change (IPCC, 2022) and may also bear CCUS costs.	Inclusive policy visions, pro-justice policy monitoring framework (Dillman and Heinonen, 2022).
High energy costs for CCUS electricity	CO_2 capture is relatively expensive (Cherepovitsyna et al., 2023).	Provide capital costs and tax incentives (Cherepovitsyna et al., 2023).
Job losses for vulnerable groups	Job losses in fossil fuel sectors are compensated by an increase in energy jobs in low-carbon sectors (Gabbatiss, 2023).	Inclusive policy visions (Dillman and Heinonen, 2022); skills set, and training requires to adapt the new emerging technologies.
Absence of marginalized communities in decision-making fora	Experiences and knowledge of more marginalized communities are not represented, recognized, or used to build more diverse ways of understanding CCUS implications (Nielsen et al., 2022).	Communication and public engagement (Brunsting et al., 2011).

roadmaps, and policies, which is in turn affected by posing questions about each of these principles as shown in Table 3.

3.3.1. *Availability*

Sovacool *et al.* (2017, p. 678) define availability as "people deserve sufficient energy resources of high quality (suitable to meet their end uses)." This principle is closely related to the notion of distributive justice. In countries such as the US, Canada, Europe, Australia, and Japan, which have several existing infrastructures and well-established CCUS projects, beginning in Texas in the early 1970s (IEAGHG, n.d.), the technology has integrated this principle. Weyburn Saskatchewan CCUS is a good example of a successful project. The Canadian government committed to adopting CCUS by launching CCUS-EOR in 2000, which conducted research and demonstration projects and built a regulatory framework, with the result that as of 2021 the project has sequestered more than 40 MtCO$_2$ (Mitrović and Malone, 2011; Government of Saskatchewan, 2021).

The availability principle must also meet the ability of economies and market capacities to provide sufficient energy (Sovacool and Dworkin, 2015). Nevertheless, the investment required to maintain system operations remains uncertain, as ASEAN continues to lack access to financing for commercialization of such technologies as of this writing. Future investments in CCUS in Southeast Asia will be reliant on the development of legislative, regulatory, and policy incentives, with foreign financing playing a significant role (IEA, 2021). Policy and innovative business models are the key drivers for CCUS investment incentives (IEF, 2021). To ensure availability, policy should prioritize developing robust and diverse CCUS value chains capable of withstanding disruptions.

3.3.2. *Affordability*

Numerous scholars state that a paramount challenge to acquiring CCUS technology is cost. "All people, including the poor, should pay no more than 10% of their income for energy services" is how Sovacool *et al.* (2017, p. 678) define affordability. Companies are more inclined to allocate resources to CCUS projects when there is a substantial infusion of funds from the government in the form of direct grant financing (GCCUSI, 2021). Such assistance is accordingly intended to bolster private sector equity investments. According to Lin and Tan's (2021) study, a high carbon price or carbon tax is a crucial prerequisite for initiating investment in CCUS. Utilizing the collected CO_2 for enhanced oil recovery has the potential to significantly boost the economic viability of such operations.

Affordability in this regard must encompass both price and price stability. The presence of available energy is meaningless when its utilization is hindered by budgetary limitations. A suitable risk allocation framework is thus necessary to support CCUS implementation, as it plays a pivotal role in generating appealing investment prospects for the private sector in ASEAN, while simultaneously fostering CCUS market growth therein. It is therefore imperative to guarantee that energy bills do not impose excessive financial strain on customers so as to establish affordable CCUS initiatives.

3.3.3. *Due process*

The concept of due process is a fundamental component of procedural justice (Sovacool and Dworkin, 2015). According to Sovacool *et al.* (2017, p. 678), this means that "countries should respect due process and human rights in their production and use of energy." The primary emphasis lies in guaranteeing stakeholder involvement in energy decision-making processes, taking steps to ensure meaningful participation and transparency throughout. Liu *et al.* (2021) assessed the risk assessment decision-making model for achieving sustainable CCUS projects. They found that economy, society, environment, governance, and technology must be considered if CCUS commercialization is to succeed. Discussions about public acceptance, information disclosure, community engagement, and outreach activities have been held for such projects as Tomakomai CCUS Project, Japan (Asia Clean Energy Forum, 2020), and CCUS projects in the UK (Traverse, 2021).

In ASEAN, Indonesia has commenced the Gundih CCUS project, initiated by Pertamina, Indonesia's state-owned oil and gas company. As the project is located in a highly populated area, a public engagement strategy of involving stakeholders was adopted to gain community acceptance. The country appeared to understand that limited knowledge of CCUS could hamper implementation. The Public Engagement Strategy that the Gundih CCUS project accordingly applied comprised (1) CCUS stakeholder analysis to understand characteristics of social sites and the level of community knowledge, (2) CCUS opinion shaping factors, i.e., stakeholders' opinions toward, and local peoples' involvement in, CCUS operations, (3) debate on CCUS pros and cons, (4) CCUS frameworks, (5) CCUS in local media, and (6) CCUS communications plans (Mulyasari *et al.*, 2021).

Limited transparency and lack of due process within the renewable energy industry have impeded the facilitation of private investor mobilization. Constrained scope for public engagement in procedures pertaining to creating renewable infrastructure gives rise to social risk, including marginalization and disempowerment of individuals intended to benefit (Setyowati, 2021). Hence, including stakeholders in energy governance and promoting fair and informed consent among impacted populations are crucial to upholding due process.

3.3.4. *Good governance*

Sovacool *et al.* (2017, p. 678) define good governance as "all people should have access to high quality information about energy and the environment and fair, transparent, and accountable forms of energy decision-making." The aspects that must be fulfilled to satisfy this principle include access to information, accountability, and transparency. Insufficient and inconsistent support for CCUS has contributed to the relatively sluggish progress in project implementation thus far.

According to research conducted by Liponnen *et al.* (2017), it was observed that interest in CCUS fluctuated during the Clean Energy Ministerial process. The study suggests that absent substantial enhancements in climate policy measures, both policy and political support for CCUS may continue to be inconsistent. As markets cannot drive CCUS implementation at the speed and scale required to fulfill climate objectives, enhanced climate response, targeted policy support, and stronger government leadership are necessary for widespread CCUS deployment. The political will and agendas of ASEAN member states play crucial roles in either hindering or facilitating such development. Therefore, in order to successfully achieve the desired outcomes, the political and institutional processes involved must take into account the aforementioned fundamental principles of good governance.

3.3.5. *Sustainability*

Over the past three decades, numerous experts in the field have forecast the necessity of CCUS for reducing carbon emissions in industries, such as energy, chemicals, cement, and steel. However, the CCUS sector has encountered difficulties in establishing a solid foundation (McKinsey, 2022). According to Sovacool *et al.* (2017, p. 678), sustainability means "energy resources should not be depleted too quickly." In the context of justice, it implies that the CCUS value chain from CO_2 capture, transportation networks, and storage infrastructure must be developed in a synchronized way.

Building CCUS projects requires that all actors involved in the CCUS supply chain, including emission sources, capture facilities, transportation systems, and storage facilities, have viable economic incentives. Supply chain establishment necessitates substantial financial commitment and continuous operational expenditures. Moreover, expenses incurred at each stage are contingent upon supply chain-specific factors that vary from project to project. Expenses associated with CO_2 capture are contingent upon CO_2 concentrations, whereas costs associated with transporting CO_2 are contingent upon factors, such as volume, distance, and topography (National Petroleum Council, 2021). Expenses associated with storage vary according to factors, such as geographical location and characteristics of storage formations. There are numerous options available when it comes to sources of CO_2, capture processes, transportation methods, and end uses (National Petroleum Council, 2021), allowing for consideration of a wide range of supply chain configurations.

There exist several potential breakthroughs that have the capacity to facilitate efficient operation of a global CO_2 value chain. The CO_2L-BLUE LCO_2 carrier developed by Mitsubishi Shipbuilding has the potential to offer a financially viable solution for addressing the increasing need to carry liquefied carbon dioxide (LCO_2) to end users or storage facilities (Wood, 2023b). A test vessel was recently inaugurated at the Shimonoseki shipyard in Japan with the purpose of developing and showcasing technology for secure and cost-effective transportation of LCO_2. Digital platforms connect many stakeholders involved in the full value chain, including CO_2 emitters, sellers, buyers, investors, LCO_2 producers, distributors, aggregators, intermediate storage providers, and sequestration providers (Wood, 2023b).

3.3.6. *Intragenerational equity and Intergenerational equity*

Sovacool *et al.* (2017, p. 678) defined intergenerational equity as "All people have a right to fairly access energy services" and intergenerational equity as "Future generations have a right to enjoy a good life undisturbed by the damage our energy systems inflict on the world today." Equitable access to energy derived from CCUS depends on the interplay of several factors, including but not limited to availability, affordability, and sustainability. Unfavorable or ambivalent public opinion toward CCUS (Shackley *et al.*, 2009), coupled with a track record of discontinued demonstration projects in several European nations, has the potential to foster skepticism toward developers of forthcoming CCUS initiatives (Boomsma *et al.*, 2020). Therefore, ensuring inter- and intragenerational equality in the context of CCUS involves guaranteeing that both current and future generations have access to a minimum level of energy services that promote their overall welfare. The extent to which this principle is adhered to may thus significantly influence future geopolitical dynamics regarding CCUS development (Dembi, 2022).

ASEAN member states have decided to commence CCUS deployment in the timeframe of 2025–2030 (ACE, 2022), meaning projects will most likely happen later rather than sooner. Hence, it would be beneficial to consider measures of intra- and intergenerational fairness when designing institutions that aim to promote interests of future generations and evaluating current CCUS progress.

3.3.7. *Responsibility*

According to Sovacool *et al.* (2017, p. 678), responsibility means that "all nations have a responsibility to protect the natural environment and minimize energy-related environmental threats." Disparities regarding environmental, health, and safety hazards associated with CCUS, as depicted in Table 4, demonstrate that the technology can engender inequality if not appropriately managed. CCUS commercialization in such aforementioned key industrial sectors as power, chemical, cement, iron, and steel is hindered by the varying technological maturity levels of its critical elements. These industries are recognized as the main emitters of CO_2 in particular while examining the integration of CO_2 capture and storage in geological formations or usage with other technologies (Dziejarski *et al.*, 2023).

CCUS initiatives encounter technical challenges, and their economic feasibility remains unknown, as seen by the unsuccessful Chevron Gorgon CCUS project, which boasted the greatest scale and had specialized geological storage in Western Australia (IEEFA, 2022a). If such adverse externalities are to be mitigated, clarification on the fundamental elements that regulate CCUS must be provided, including factors such as application and authorization processes, and necessary technical and environmental management requirements.

4. Conclusion

The investigation of energy justice in the context of CCUS highlights the need for ASEAN member states to incorporate energy justice principles in CCUS development. The concept of energy justice pertains to a global society that ensures fair distribution of advantages and disadvantages associated with energy production and consumption, ensuring that everyone has the opportunity to participate in decision-making processes concerning the development of energy systems.

The viability of CCUS as a solution for mitigating climate change has faced criticism due to sluggish development, associated hazards, and utilization of such technologies as improved oil recovery that might potentially contribute to increased carbon emissions. During CCUS implementation, potential injustices may occur in the form of procedural justice concerns, including lack of public acceptability, entrenched political and economic interests, and absence of well-defined vision. The potential for injustice in distributive justice arises from such factors as elevated costs associated with power generation through CCUS as well as attendant safety hazards, health and environmental dangers, inadequate financial stability, and dependence on gas and coal infrastructure. Finally, with regard to recognizing justice, issues encompass lack of community involvement in decision-making processes, unemployment, and climate change discrimination.

The energy justice framework principles are employed to discern instances of inequity by addressing where injustices occur, who experiences their impacts, how they are affected, and what measures may be taken to address these issues. Upon analyzing the aforementioned issues via the lens of justice principles, it is concluded that facilitating commercial-scale adoption of CCUS in ASEAN member states can be achieved through implementing said principles. This entails identifying areas characterized by injustice and implementing suitable mitigations.

References

Arlota, C. and Costa, H. K. D. M. (2021). Climate change, carbon capture and storage (CCS), energy transition, and justice: Where we are now, and where are (should be) we headed? *Carbon Capture and Storage in International Energy Policy and Law*, 385–393. https://doi.org/10.1016/b978-0-323-85250-0.00019-0.

Batres, M., Wang, F. M., Buck, H., Kapila, R., Kosar, U., Licker, R., Nagabhushan, D., Rekhelman, E., and Suarez, V. (2021). Environmental and climate justice and technological carbon removal. *The Electricity Journal*, *34*(7), 107002. https://doi.org/10.1016/j.tej.2021.107002.

Boomsma, C., ter Mors, E., Jack, C., Broecks, K., Buzoianu, C., Cismaru, D. M., Peuchen, R., Piek, P., Schumann, D., Shackley, S., and Werker, J. (2020). Community compensation in the context of carbon capture and storage: Current debates and practices. *International Journal of Greenhouse Gas Control*, *101*, 103128. https://doi.org/10.1016/j.ijggc.2020.103128.

Brunsting, S., Upham, P., Dütschke, E., De Best Waldhober, M., Oltra, C., Desbarats, J., Riesch, H., and Reiner, D. (2011). Communicating CCS: Applying communications theory to public perceptions of carbon capture and storage. *International Journal of Greenhouse Gas Control*, *5*(6), 1651–1662. https://doi.org/10.1016/j.ijggc.2011.09.012.

Bui, M., Adjiman, C. S., Bardow, A., Anthony, E. J., Boston, A., Brown, S., Fennell, P. S., Fuss, S., Galindo, A., Hackett, L. A., Hallett, J. P., Herzog, H. J., Jackson, G., Kemper, J., Krevor, S., Maitland, G. C., Matuszewski, M., Metcalfe, I. S., Petit, C., Puxty, G., Reimer, J., Reiner, D. M., Rubin, E. S., Scott, S. A., Shah, N., Smit, B., Trusler, J. P. M., Webley, P., Wilcox, J., and Mac Dowell, N. (2018). Carbon capture and storage (CCS): The way forward. *Energy & Environmental Science*, *11*(5), 1062–1176. https://doi.org/10.1039/c7ee02342a.

CCUS Investor Roadmap — Policies — IEA. (n.d.). *IEA*. https://www.iea.org/policies/16834-ccus-investor-roadmap.

Chen, S., Liu, J., Zhang, Q., Teng, F., and McLellan, B. C. (2022). A critical review on deployment planning and risk analysis of carbon capture, utilization, and storage (CCUS) toward carbon neutrality. *Renewable and Sustainable Energy Reviews*, *167*, 112537. https://doi.org/10.1016/j.rser.2022.112537.

Cherepovitsyna, A., Kuznetsova, E., and Guseva, T. (2023). The costs of CC(U)S adaptation: The case of Russian power industry. *Energy Reports*, *9*, 704–710. https://doi.org/10.1016/j.egyr.2022.11.104.

Costley, D. (2022). Battle over carbon capture as tool to fight climate change. *AP News*. https://apnews.com/article/climate-technology-science-business-louisiana-8e41ce52bcb66117ad97b58dffe3b607.

Dembi, V. (2022). Ensuring energy justice in transition to green hydrogen. Social Science Research Network; RELX Group, Netherlands. https://doi.org/10.2139/ssrn.4015169.

Dillman, K. and Heinonen, J. (2022). A 'just' hydrogen economy: A normative energy justice assessment of the hydrogen economy. *Renewable and*

Sustainable Energy Reviews, *167*, 112648. https://doi.org/10.1016/j.rser. 2022.112648.

Dziejarski, B., Krzyżyńska, R., and Andersson, K. (2023). Current status of carbon capture, utilization, and storage technologies in the global economy: A survey of technical assessment. *Fuel*, *342*, 127776. https://doi.org/10.1016/j.fuel.2023.127776.

Fraser N. (1999). Social justice in the age of identity politics. In Henderson, G. (ed.) *Geographical Thought: A Praxis Perspective*. London: Taylor and Francis, pp. 56–89.

Gabbatiss, J. (2023). Jobs created by net-zero transition will 'offset' fossil-fuel job losses in Republican US states. *Carbon Brief.* https://www.carbonbrief. org/jobs-created-by-net-zero-transition-will-offset-fossil-fuel-job-losses-in-republican-us-states/.

GCCSI. (2021). Financing CCS in developing countries prepared for clear path and southern company. https://www.globalccsinstitute.com/wp-content/ uploads/2021/04/Financing-CCS-In-Developing-Countries-V2-1.pdf. Accessed date: August 13, 2023.

Ghilloti. (2022). Latest oil and gas news. *Upstream Online*. Retrieved August 6, 2023, from https://www.upstreamonline.com/energy-transition/high-carbon-prices-spurring-europe-s-ccs-drive/2-1-1308488.

Government of Saskatchewan. (n.d.). Saskatchewan announces carbon capture utilization and storage priorities. Government of Saskatchewan. https:// www.saskatchewan.ca/government/news-and-media/2021/september/07/ saskatchewan-announces-carbon-capture-utilization-and-storage-priorities.

IEA. (2019). Global CO_2 emissions in 2019. https://www.iea.org/articles/ global-co2-emissions-in-2019.

IEA. (2021a). Carbon capture, utilisation and storage: The opportunity in Southeast Asia. https://www.iea.org/reports/carbon-capture-utilisation-and-storage-the-opportunity-in-southeast-asia.

IEA. (2021b). Net zero by 2050: A roadmap for the global energy sector. https:// www.iea.org/reports/net-zero-by-2050.

IEA. (2022). Carbon capture, utilisation and storage. International Energy Agency, Paris. https://www.iea.org/reports/carbon-capture-utilisation-and-storage-2.

IEAGHG. (n.d). A brief history of CCS and current status. https://ieaghg.org/ docs/General_Docs/Publications/Information_Sheets_for_CCS_2.pdf.

IEEFA. (2022a). 'Carbon capture' model at Exxon's Shute Creek CCUS reveals a questionable technology and uncertain economic viability. https:// ieefa.org/resources/carbon-capture-model-exxons-shute-creek-ccus-reveals-questionable-technology-and. Accessed date: August 7, 2023.

IEEFA. (2022b). Carbon capture in the Southeast Asia. https://ieefa.org/wp-content/uploads/2022/04/Carbon-Capture-in-the-Southeast-Asian-Market-Context_April-2022.pdf. Accessed date: August 7, 2023.

IEEFA. (2022c). The carbon capture crux: Lessons learned. https://ieefa.org/resources/carbon-capture-crux-lessons-learned. Accessed date: August 13, 2023.

IEF. (2021). Strategies to scale carbon capture utilization and storage. *Dialogue Insight Report*, September 2021. International Energy Forum. https://www.ief.org/_resources/files/events/strategies-to-scale-carbon-capture-utilization-and-storage/ccus-report.pdf.

IPCC. (2005). Bert Metz, Ogunlade Davidson, Heleen de Coninck, Manuela Loos and Leo Meyer (Eds.) Cambridge University Press, UK. pp 431. Available from Cambridge University Press, The Edinburgh Building Shaftesbury Road, Cambridge CB2 2RU ENGLAND. https://www.ipcc.ch/report/carbon-dioxide-capture-and-storage/.

IPCC. (2022). Climate change 2022: Impacts, adaptation and vulnerability. https://www.ipcc.ch/report/ar6/wg2/downloads/report/IPCC_AR6_WGII_SummaryVolume.pdf.

Jenkins, K., McCauley, D., Heffron, R., Stephan, H., and Rehner, R. (2016). Energy justice: A conceptual review. *Energy Research & Social Science, 11*, 174–182. https://doi.org/10.1016/j.erss.2015.10.004.

Lin, B. and Tan, Z. (2021). How much impact will low oil price and carbon trading mechanism have on the value of carbon capture utilization and storage (CCUS) project? Analysis based on real option method. *Journal of Cleaner Production, 298*, 126768. https://doi.org/10.1016/j.jclepro.2021.126768.

Lipponen, J., McCulloch, S., Keeling, S., Stanley, T., Berghout, N., and Berly, T. (2017). The politics of large-scale CCS deployment. *Energy Procedia, 114*, 7581–7595. https://doi.org/10.1016/j.egypro.2017.03.1890.

Liu, B., Liu, S., Xue, B., Lu, S., and Yang, Y. (2021). Formalizing an integrated decision-making model for the risk assessment of carbon capture, utilization, and storage projects: From a sustainability perspective. *Applied Energy, 303*, 117624. https://doi.org/10.1016/j.apenergy.2021.117624.

McCauley, D. A., Heffron, R. J., Stephan, H., and Jenkins, K. (2013). Advancing energy justice: The triumvirate of tenets. *International Energy Law Review, 32*(3), 107–110. http://hdl.handle.net/1893/18349.

Mitrović, M. and Malone, A. (2011). Carbon capture and storage (CCS) demonstration projects in Canada. *Energy Procedia, 4*, 5685–5691. https://doi.org/10.1016/j.egypro.2011.02.562.

Moss, J. (2009). *Climate Change and Social Justice*. Melbourne: Melbourne University Publishing, pp. 51–67.

Nielsen, J. A. E., Stavrianakis, K., and Morrison, Z. (2022, August 2). Community acceptance and social impacts of carbon capture, utilization and storage projects: A systematic meta-narrative literature review. *PLOS One, 17*(8), e0272409. https://doi.org/10.1371/journal.pone.0272409.

Owens, S. and Driffill, L. (2008). How to change attitudes and behaviours in the context of energy. *Energy Policy*, *36*(12), 4412–4418. https://doi. org/10.1016/j.enpol.2008.09.031.

Preston, C. and Carr, W. (2018*).* Recognitional justice, climate engineering, and the care approach. *Ethics, Policy & Environment, 21*(3), 308–323. https:// doi.org/10.1080/21550085.2018.1562527.

Sawada, Y. (2020). Experience of Tomakomai CCS Project. *Asia Clean Energy Forum 2020.* https://asiacleanenergyforum.adb.org/wp-content/uploads/ 2020/07/yoshihiro-sawada-experience-of-tomakomai-ccs-project.pdf.

Setyowati, A. B. (2021). Mitigating inequality with emissions? Exploring energy justice and financing transitions to low carbon energy in Indonesia. *Energy Research & Social Science, 71,* 101817. https://doi.org/10.1016/ j.erss.2020.101817.

Sovacool, B. K. and Dworkin, M. H. (2015). Energy justice: Conceptual insights and practical applications. *Applied Energy, 142,* 435–444. https://doi. org/10.1016/j.apenergy.2015.01.002.

Sovacool, B. K., Burke, M., Baker, L., Kotikalapudi, C. K., and Wlokas, H. (2017). New frontiers and conceptual frameworks for energy justice. *Energy Policy, 105,* 677–691. https://doi.org/10.1016/j.enpol.2017.03.005.

Sovacool, B. K., Martiskainen, M., Hook, A., and Baker, L. (2019). Decarbonization and its discontents: A critical energy justice perspective on four low-carbon transitions. *Climatic Change* (Springer Science+Business Media). https:// doi.org/10.1007/s10584-019-02521-7.

Stake, R. (1995). *The Art of Case Study Research.* Thousand Oaks, CA: Sage Publications.

Storrs, K., Lyhne, I., and Drustrup, R. (2023). A comprehensive framework for feasibility of CCUS deployment: A meta-review of literature on factors impacting CCUS deployment. *International Journal of Greenhouse Gas Control, 125,* 103878. https://doi.org/10.1016/j.ijggc.2023.103878.

Taylor, K. and Taylor, K. (2022). EU energy chief announces "strategic vision" for CCUS in 2023. www.euractiv.com. https://www.euractiv.com/section/ energy-environment/news/eu-energy-chief-announces-strategic-vision-for-ccus-in-2023/.

ter Mors, E. and van Leeuwen, E. (2023). It matters to be heard: Increasing the citizen acceptance of low-carbon technologies in the Netherlands and United Kingdom. *Energy Research & Social Science, 100,* 103103. https://doi. org/10.1016/j.erss.2023.103103.

Traverse. (2021). Carbon capture usage and storage public dialogue. *Department for Business, Energy, & Industrial Strategy.* https://assets.publishing.service. gov.uk/government/uploads/system/uploads/attachment_data/file/1005434/ ccus-public-perceptions-traverse-report.pdf.

Wallsgrove, R., Woo, J., Lee, J. H., and Akiba, L. (2021). The emerging potential of microgrids in the transition to 100% renewable energy systems. *Energies, 14*(6), 1687. https://doi.org/10.3390/en14061687.

Wang, N., Akimoto, K., and Nemet, G. F. (2021). What went wrong? Learning from three decades of carbon capture, utilization and sequestration (CCUS) pilot and demonstration projects. *Energy Policy, 158*, 112546. https://doi.org/10.1016/j.enpol.2021.112546.

Wei, N., Li, X., Liu, S., Lu, S., and Jiao, Z. (2021). A strategic framework for commercialization of carbon capture, geological utilization, and storage technology in China. *International Journal of Greenhouse Gas Control, 110*, 103420. https://doi.org/10.1016/j.ijggc.2021.103420.

Wood, N. (2023a). Problematising energy justice: Towards conceptual and normative alignment. *Energy Research & Social Science, 97*, 102993. https://doi.org/10.1016/j.erss.2023.102993.

Wood, J. (2023b). What is the CO_2 value chain and why is it key for net zero? *Spectra*. Retrieved August 13, 2023. https://spectra.mhi.com/what-is-the-co2-value-chain-and-why-is-it-key-for-net-zero.

Chapter 3

Carbon Capture, Usage, and Storage (CCUS) Implications in Thailand

Weerawat Chantanakome

Ministry of Energy, Bangkok, Thailand

weerawat.wc@gmail.com

Abstract

Escalating global energy demand, largely met through fossil fuel consumption, has driven surging carbon emissions, posing a substantial threat to climate change. Thailand has committed to achieving net zero emissions by 2065 and carbon neutrality by 2050 as part of the Paris Agreement and the 26th annual session of the Conference of the Parties (COP 26) in 2021. Various policies and mitigation options have been implemented in Thailand, including adopting electric vehicles, biofuels, green building practices, energy conservation, and renewable energy sources. This chapter mainly focuses on carbon capture, usage, and storage (CCUS) technology and explores its potential as a critical decarbonization tool to address Thailand's increasing energy demand and greenhouse gas emissions. Successful deployment requires that government and industry collaborate to establish a sustainable market and achieve cost reductions. It concludes that scaling up CCUS technologies and fostering CCUS collaborations are crucial for achieving a cleaner, sustainable energy future.

Keywords: Carbon emissions, carbon capture, carbon utilization and storage, energy transition, CCUS project development in Thailand.

1. Introduction

The modern world's rapidly expanding energy demand, necessitating deriving energy from fossil fuel burning as the world's principal energy source, has accelerated the rate of carbon emissions in the atmosphere. This poses a significant threat to global climate change, with warming and other environmental repercussions readily evident. Fossil fuel burning in the power generation, transportation, and industrial sectors has helped drive the world's energy-related carbon dioxide emissions to reach 36.3 BtonCO$_2$eq in 2021, a record high. The 6 percent increase was attributed to the global economy's robust recovery from the COVID-19 pandemic and its heavy reliance on coal. Efforts to address climate change must include a focus on East Asia and the Pacific, which is responsible for one-third of the world's greenhouse gas emissions (GHGs) and 60% of its coal burning.

Thailand's emissions increased 2.29 percent annually between 1990 and 2021, from 86.71 MtonCO$_2$eq to 278.50 MtonCO$_2$eq, owing to economic and population growth. Thailand's overall emissions in 2019 were 290.24 MtonCO$_2$eq, of which the industrial sector was responsible for 17.21 percent. As a result, the Paris Agreement goal of keeping global temperature rise to 2°C, or ideally 1.5°C, by the end of the century remains in place. Thailand has also pledged to attain net zero emissions by 2065 and carbon neutrality by 2050, in accordance with the Conference of the Parties' 26th annual session (COP 26) in Glasgow, UK, in 2021.

Thailand has accordingly adopted many policies and mitigation options to reduce its emissions. Many technologies have been used for short-, medium-, and long-term mitigation, such as electric vehicles and biofuel in the transport sector, green building and energy conservation in the residential sector, and renewables in the energy sector. Thailand is also investigating the current situation and comprehensive viewpoints of Carbon Capture Utilization and Storage (CCUS) as an important decarbonization technology as a consequence of the country's rising energy demand and GHGs.

The next few decades may be challenging to prospects for such technologies as CCUS as well as for putting the global energy system on a net-zero path. Commercializing prototype CCUS seems necessary to provide motivation for the enhancement of and cost reductions in decarbonization technology. While government policy has a crucial role in creating a sustained and viable market for CCUS, the industry must also embrace the opportunity.

Markets alone cannot make CCUS a clean energy success story. Government and industry should have the chance to pool their resources in terms of expertise and financial skills if we are to enjoy the environmental and economic benefits that CCUS offers. Without it, achieving our energy and climate goals will be next to impossible.

CCUS participation in the energy sector comes after acknowledging the energy transitions that have made our energy systems more secure, affordable, and sustainable. At the same time, we recognize that there are still several national paths that can be taken, depending on the conditions in each country and the level of public and private sector support for the CCUS goal.

1.1. *CCUS technology: Overview*

Figure 1 shows the overall CCUS chain. The first phase involves capturing carbon emissions from different sectors, as these emissions are not limited to power production. Other sectors such as heavy industry,

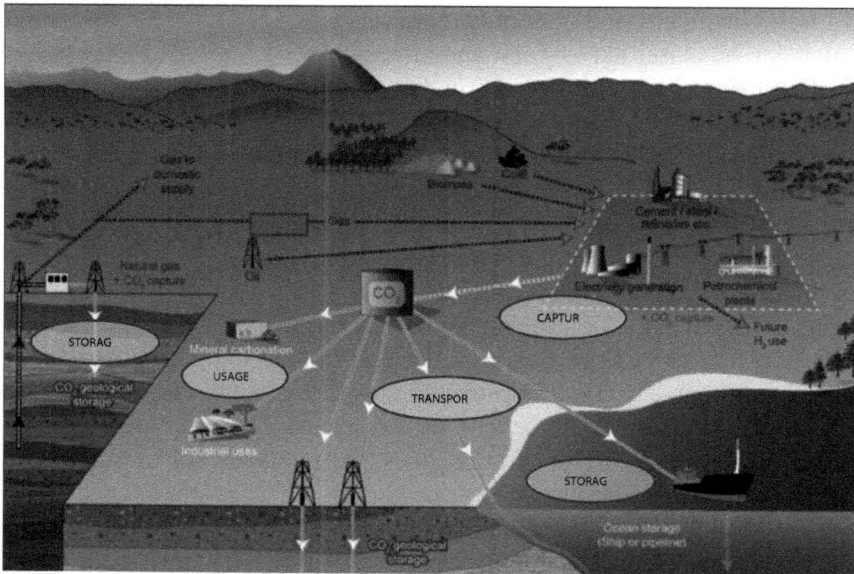

Figure 1.   Carbon capture, utilization, and storage.

Source: https://archive.ipcc.ch/publications_and_data/ar4/wg3/en/figure-4-22.html.

i.e., cement and ceramic plants, transportation, waste disposal, and chemical treatment plants also emit GHGs. Carbon capture technology involves absorption, adsorption, membranes, and high-temperature lopping (Carbon Capture Utilisation and Storage in the European Union – 2022). This classification mainly applies to carbon capture in power generation facilities, namely pre, post, and oxy-combustion. In other industries, however, which emit carbon during processing rather than burning fuel, carbon capture is instead categorized based on the carbon separation process (Carbon Capture Utilization and Storage — Publications Office of the EU, n.d.). Once captured, CO_2 is compressed to a supercritical state, transported, injected, and/or used. The compression step is usually included in the capture system.

Next is the carbon utilization phase, chemically transforming CO_2 into another product with commercial value. CO_2 utilization has attracted interest due to its potential to replace non-sustainable fossil fuels by recycled CO_2 that could both avoid burning fossil fuels and net atmospheric CO_2 emissions (Carbon Capture Utilisation and Storage — Publications Office of the EU, n.d.). It has also emerged as a source of potential competitive advantage for European industry in the production of fuels, chemicals, and materials. A variety of CO_2 sources is available, which can be classified as point and atmospheric.

Following capture and/or use, in the third phase CO_2 is transported via pipelines and/or shipped for injection, i.e., to storage sites, where it is stored in deep saline aquifers, deep coal bed methane, i.e., enhanced, combined, or used in enhanced oil recovery (EOR), in depleted oil/gas reservoirs, and most recently in basalt. It is then monitored for safety by accurate geochemical and geophysical technologies.

Figure 2 explains the stages, activities, and future opportunities for decarbonization in CCUS. The scope of this CCUS study is divided into two parts. The first scope is the transportation and storage study; it is the Carbon capture and storage (CCS) for upstream industries, which is covered in Phase One. The components of Phase One for the upstream industry are economic and incentive policy, technical guidelines and regulation of transportation and storage, and partnerships. This scope covers offshore sources and sinks, domestic shipping, and onshore sinks for transportation and storage.

The second scope is the capture study; it is CCS for downstream industry and CO_2 capture and collection from industrial sources and power generation, which are covered in scopes 2 and 3. It also covers collection terminals and onshore sources.

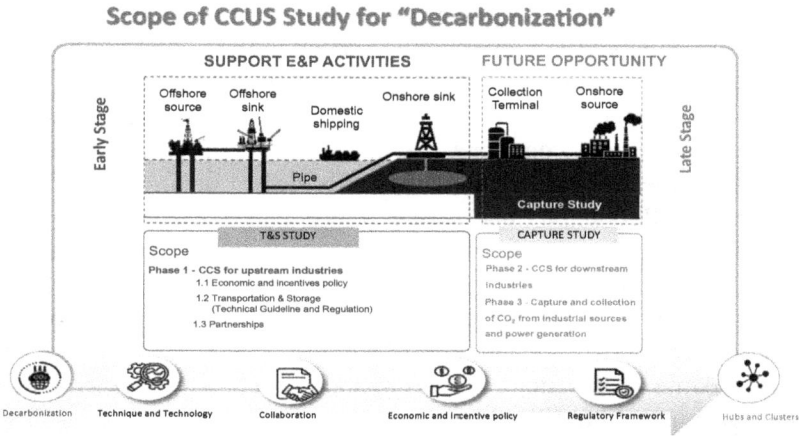

Figure 2.   Scope of CCUS study for decarbonization.

2. Overall CCUS Policy Implications

At present, knowledge of CO_2 storage is still expanding. Numerous nations have enhanced their knowledge of domestic and subsurface CO_2 storage capacity, leading to increases in overall results. The top three countries in terms of scores and readiness for extensive deployment are Canada, Norway, and the US. While there is much optimism about the potential of global storage resources after the addition of the following nine high-scoring countries, however, there is a global slowdown in the development of storage resources and site characterization. Norway tops the list by a wide margin. Its ranking is evidence of its government and business sectors' ongoing dedication to CCS, as emphasized by recent government announcements. Research, project feasibility studies, carbon fees, and other crucial supportive measures are all currently being implemented. Following Norway in Band A are the US, China, Canada, the UK, and Japan. Strong evidence exists that these countries are dedicated to CCS. China has the most large-scale CCS facilities under development and in various stages of planning thanks to supportive policy. Therefore, deployment is strongly influenced by policy, but more action is needed.

However, 27 of the 135 CCUS initiatives in development are already up and running. In the first 9 months of 2021, 71 new projects were introduced. In each location, there have been 36 initiatives started, including 8 in the US, 5 in the UK, 4 in the Netherlands, and 36 in Belgium.

In Asia, positive strides have been achieved over the past year, with the first commercial CCS projects announced in Indonesia and Malaysia, despite the fact that their CCUS investments lag behind those in Europe and the US. Systems for the economy are also being created. The first financial incentive system for CCS in the Asia-Pacific area was established when the Australian government decided to incorporate CCS into the Emissions Reduction Fund (ERF). Additionally, 2,225 power plants in China have started an emissions trading program that will exchange more than 4 $GtCO_2$ annually.

Although CCUS is more expensive than other GHG reduction technologies, it can store huge amounts of carbon. It also contributes to stable soil granulation, has better ventilation, and has a better capacity to retain water, affecting soil fertility and plant nutrients. This feature makes it suitable for agriculture, which is the main industry in Thailand.

2.1. *Worldwide CCS and CCUS development*

Carbon capture and storage (CCS or CCUS) is an essential and urgently available technology to mitigate the effects of climate change. In terms of internationally recognizing the need for explicit action to enhance the world's ability to address the climate crisis, broader use of CCS will support significant reductions in GHGs (Bhatia *et al.*, 2019). However, despite being proven successful and accessible, CCS is expensive. As a result, there is still a shortage of investment in this field.

CCS manages CO_2 emissions that are collected from various processes, such as oxy-fuel combustion, post-combustion, and pre-combustion. There are four stages to a CCS project: CO_2 capture, CO_2 transportation, CO_2 injection, and CO_2 post-injection. CCUS has gained popularity in recent years since it can lower sequestration costs and provide benefits by increasing the production of hydrocarbons or geothermal energy (Bajpai *et al.*, 2022). In accordance with the goal of CO_2 injection, a number of related technologies have been created, including EOR, enhanced coalbed methane recovery (ECBM), enhanced gas recovery (EGR), enhanced shale gas recovery (ESG), and enhanced geothermal system (EGS).

Figure 3 shows CO_2 emissions reductions in the energy sector under different Sustainable Development Scenarios (SDS), for 2022, 2030, 2050, and 2065. CCS can contribute more to carbon reduction as it matures and becomes cost-effective. In 2022, the contribution of CCS

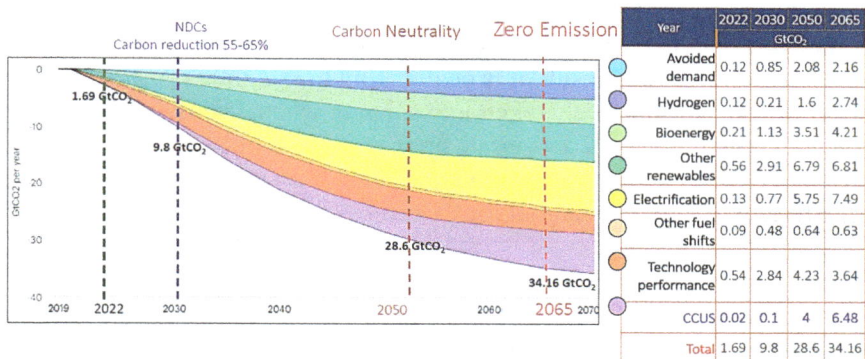

Year	2022	2030	2050	2065
	GtCO$_2$			
Avoided demand	0.12	0.85	2.08	2.16
Hydrogen	0.12	0.21	1.6	2.74
Bioenergy	0.21	1.13	3.51	4.21
Other renewables	0.56	2.91	6.79	6.81
Electrification	0.13	0.77	5.75	7.49
Other fuel shifts	0.09	0.48	0.64	0.63
Technology performance	0.54	2.84	4.23	3.64
CCUS	0.02	0.1	4	6.48
Total	1.69	9.8	28.6	34.16

Figure 3. CO$_2$ emissions reductions in the energy sector in sustainable development scenarios.

Source: https://www.iea.org/reports/ccus-in-clean-energy-transitions/ccus-in-the-transition-to-net-zero-emissions.

technology in carbon reduction was only 0.02 GtCO$_2$, while in 2065, it could potentially reduce atmospheric GHGs by 6.48 GtCO2. Thus, CCS technology contributes 18.96 percent in achieving the 2065 net zero emission scenario, more than all other technologies (IEA, 2020).

Both CCS and CCUS engineering projects are systematically difficult, and their success depends on thorough research in a variety of technical and scientific fields, including geology, geoengineering, geophysics, environmental engineering, mathematics, and computer science. Additionally, choosing the right location for any such project necessitates careful consideration of safety, economy, the environment, and public opinion at all levels of operation, including at the national, basin-wide, regional, or sub-basin levels.

CCS and CCUS are practical solutions for drastically lowering CO$_2$ emissions from large-scale emission sources. Storage locations may include deep saline formations, deep unmendable coal seams, exhausted oil or gas reservoirs, and rock salt caverns when their only function is to store CO$_2$. It is still too expensive for widespread commercial use.

Planning for CCUS systems has as its main goal enhanced usage of captured CO$_2$ to combat climate change. It is therefore essential to design an incentive program due to the energy and financial costs of CO$_2$ capture or a system for pricing carbon to promote implementation (Lambert *et al.*, 2016; Li *et al.*, 2015, 2016). On the other hand, the expansion of alternatives for CO$_2$ storage or utilization may hasten CCUS adoption for climate change mitigation. However, the use of geological sequestration for carbon management is constrained by public perception and a lack of regulatory frameworks. As a result, it is challenging to integrate use and storage because utilization is rarely taken into account when discussing CO_2 storage. Planning tools that concentrate on these challenges seek to maximize the application of both techniques nationally or internationally. These initiatives point to the potential for CCU and CCS to be combined to form CCUS.

3. Thailand's CCUS Landscape

At COP 26, on November 1, 2021, the Prime Minister of Thailand announced that Thailand would increase its NDC plan to 40 percent from BAU to reach carbon neutrality and Net Zero emissions (NZE) in 2050 and 2065, respectively (Energy Policy and Planing Office (EPPO), Ministry of Energy, 2021). Thus, in the period 2030–2065, the energy sector will be key to mitigating GHG emissions in the Long-term Low Greenhouse Gas Emission Development strategy (LT-LEDs). Specifically, in the energy system, there are several technologies that could help to reduce GHG emissions in the country, as shown in Figure 4. First, power generation will be developed using energy efficiency improvements such as natural gas with CCS and increasing the share of renewable electricity in total electricity generation to 33 percent. Second, energy efficiency improvement in other sectors including transport, residential, industry, and commercial could also decrease GHG emissions, for example, by increasing the proportion of new efficient vehicle fleets, such as PEVs (UN Global Compact, 2020).

3.1. *Maturity of CCUS technology for Thailand's green growth*

Key actions towards 2050 CN and 2065 Net-Zero are defined depending on the maturity of the industry and technologies. Such maturity consists of (1) planning and strategy formulation, (2) support for technology development, (3) policy development and enhancement, and (4) market creation for commercialization, with the last of these representing the highest maturity level. Providing incentives creates a market to revitalize business for commercialization, and attracts foreign direct investment to further accelerate the same. The policy scope of the measures for each key area herein will focus on EV and batteries, solar, and energy efficiency.

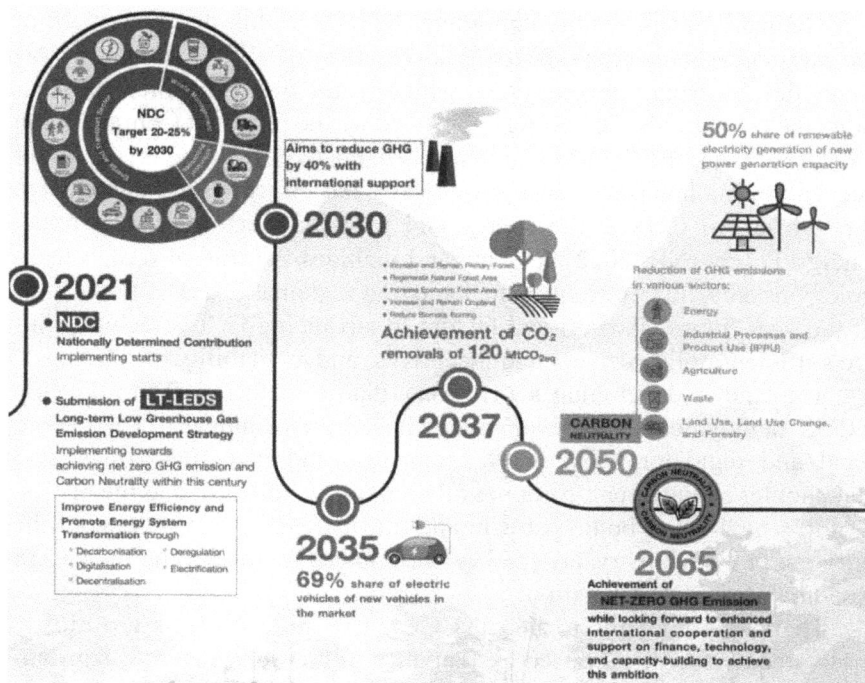

Figure 4. NDC with LT-LEDs in 2030.
Source: https://globalcompact-th.com/news/detail/602.

Policy development and enhancement promote applying technologies for commercialization and liberalize the use of technology to cut costs. The policy scope herein will also focus on EV and batteries, solar, and energy efficiency, as well as ESS/smart grid and carbon storage. Support for technology development offers government assistance, including financial support, for technology development and demonstration and invites players with leading technologies to participate in technology transfers. The policy scope herein covers the key areas of ESS/smart grid, carbon storage, carbon utilization, hydrogen and ammonia, and energy efficiency. Planning and strategy formulation are the industry or technology's lowest maturity level. It assesses the potential and relevance of the industry for the country, formulates national strategy, and sets goals and directions. It also engages in dialogue with stakeholders in order to align with international standards. Its policy scopes are the key areas of carbon storage, carbon utilization, hydrogen and ammonia, and carbon pricing.

Thailand recognizes the importance of CCS within the framework of its own decarbonization efforts. The creation of the Greenhouse Gas Reduction Steering Subcommittee, which would direct technical applications for the nation's CCUS, was approved by the National Committee on Climate Change Policy at its meeting in January 2022. The mission will accelerate Thailand's response to reducing the effects of climate change through use of CUS in the energy and industrial sectors. The National Energy Policy (NEP2022) for Thailand includes the role of CCUS technology as a useful contribution to achieving sustainable use of fossil fuels in the near future. Two critical factors in advancing CCUS technologies are sufficient policy drivers and incentives, and availability of funding.

Thailand is developing a Five Year Plan for 2022–2027 to develop CCUS deployment and development based on technical development, legal and regulatory framework, economic and incentive policy, and stakeholder engagement. National oil companies, utilities, and the private sector, which cover both upstream and downstream businesses under the purview of the Ministry, are the key participants in developing the CCUS Roadmap for decarbonization.

Thailand's attempts to install CCUS trial pilot projects are mostly motivated by the threat posed by rising CO_2 emissions. There is substantial domestic demand for CCUS because of this potential CO_2 source, including CCUS installations retrofitted in already-existing coal-fired power plants and prospective CO_2 sinks from depleted reservoirs. As of this writing, the government is preparing a CCUS business and revenue model, which it anticipates completing by 2026. The majority of CCUS initiatives are presently in the sandbox stage.

Existing power plants with CCS and such alternative technologies as energy storage, clean hydrogen, fuel ammonia, and smart grid, will be developed and integrated to handle intermittency generation from RE sources that can ensure energy security in the electricity network. CCUS will therefore help achieve sustainable use of fossil fuels in the region, especially given the current scenario. Joint efforts are needed to achieve a just and realistic energy transition while taking CCUS into account for the long-term regional energy agenda. Figure 5 shows the planned CO_2 Emission from Energy Consumption in Thailand by sector in the period 2020–2065. Note the potential of CCS technology in reducing carbon emissions in Thailand, beginning in 2045. CCS will accordingly enable Thailand to reduce carbon emissions by 55 percent in 2065. Thailand has a theoretical storage capacity of 2.69 $GtCO_2$.

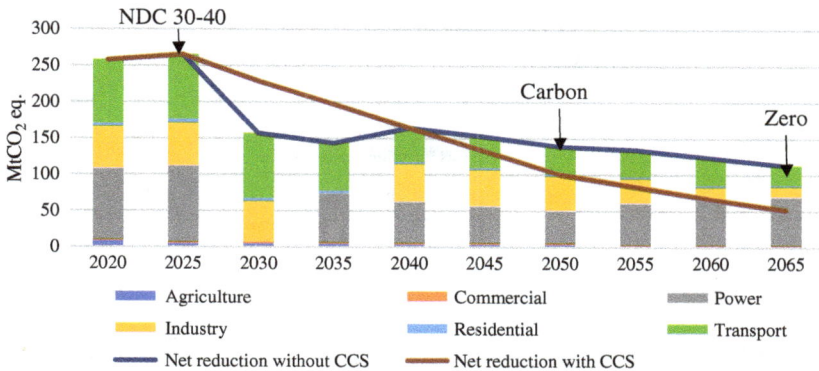

Figure 5. Potential of CCS to reduce Thai carbon emissions.

Meeting national and worldwide climate and energy goals also requires strengthened international collaboration on CCUS. The Asia CCUS Network (ACN) was accordingly established as a platform for discussion of policy, technologies, legal concerns, and supporting measures for CCUS development and deployment throughout Asia.

In a nutshell, the key to success will be a close-knit network of professionals capable of affordably commercializing these technologies. Efforts to include CCUS in our long-term regional energy roadmap coincide with the release of this report.

3.2. CCUS project developments in Thailand

Thailand has begun research into carbon storage capacity, which is still in the early stages, with no demonstration projects underway, while the technology is already established and implemented overseas. However, key sections of carbon storage for 2025 are required to formulate the future plan while promoting public acceptance. It also starts a feasibility study for the demonstration in different locations and initiates coordination with stakeholders.

Thailand has decided that carbon utilization will be a key area, requiring further discussion. The key actions towards 2025 of carbon utilization are studying carbon utilization potential in terms of CO_2 reduction impact and value of the material produced, and deciding on prioritized areas.

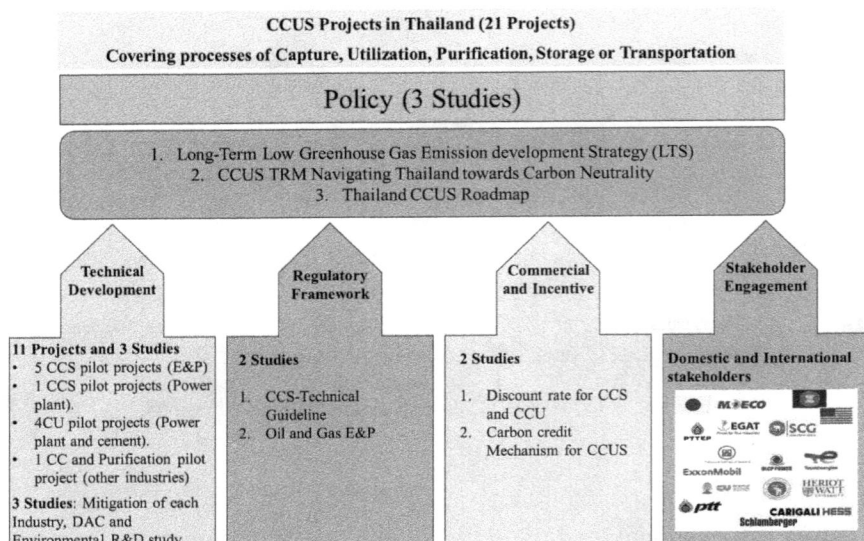

CCUS Projects in Thailand (21 Projects)
Covering processes of Capture, Utilization, Purification, Storage or Transportation

Policy (3 Studies)

1. Long-Term Low Greenhouse Gas Emission development Strategy (LTS)
2. CCUS TRM Navigating Thailand towards Carbon Neutrality
3. Thailand CCUS Roadmap

Technical Development	Regulatory Framework	Commercial and Incentive	Stakeholder Engagement
11 Projects and 3 Studies • 5 CCS pilot projects (E&P) • 1 CCS pilot projects (Power plant). • 4CU pilot projects (Power plant and cement). • 1 CC and Purification pilot project (other industries) **3 Studies**: Mitigation of each Industry, DAC and Environmental R&D study	**2 Studies** 1. CCS-Technical Guideline 2. Oil and Gas E&P	**2 Studies** 1. Discount rate for CCS and CCU 2. Carbon credit Mechanism for CCUS	**Domestic and International stakeholders**

Figure 6.   Framework for overall CCUS development in Thailand.

The Department of Mineral Fuels (DMF) presented a five year plan for 2022–2027 to develop CCUS based on a strategy including technical development, legal and regulatory framework, commercial and incentive policy, and stakeholder engagement as mentioned above and shown in Figure 6. Technical development involves strategic planning and integration of CCUS in various areas such as power plants and cement plants, and other studies focusing on emission mitigation, direct air capture (DAC), and research. Regulatory frameworks deal with the technical guidelines to handle the technology. Rebates and incentives for CCUS development and carbon market development come next, followed by engaging domestic and international stakeholders to share resources and manpower in CCUS development. The details of the CCUS projects in Thailand are shown in Table 1. Storage bodies, i.e., reservoirs, have been identified for Project 1, the Arthit Pilot Project, where PTT supports in terms of maximizing CO_2 removal. Project 2, the Kra Basin Pilot Project, and Project 3, the Phu Horm Pilot Project, are still in the stage of obtaining theoretical storage capacities. Project 4, the S1 Pilot Project, also includes EOR evaluation, and EGAT and PTTEP are discussing potential collaboration.

Table 1. CCUS pilot projects in Thailand.

Pilot projects	Oil and gas reservoir		Timeline 2022 2023 2024 2025 2026	Expected outcomes	Reservoir dept (m)
T1. Arthit Pilot Project	Gas	Sandstone (Offshore)	←————————★	Proof of concept: Injectivity and storage in different reservoirs onshore and offshore	1,500–2,500
T2. Kra Basin Pilot Project	Oil	Sandstone (Nearshore)	↕		1,000–2,500
T3. Phu Horm Pilot Project	Gas	Carbonate (Onshore)	↕		3,000–4,000
T4. S1 Pilot Project	Oil	Sandstone (Onshore)	↕		1,500–2,000
T5. BLCP Pilot Project		Coal-Fired Power plant	↕	Support CO_2 Utilization and CCUS facility retrofitted in existing coal-fired power plant	Coal-fired power plant 1. Methane Production 2. Algae cultivate

Project 5, the BLCP Pilot Project, is still in the initial phase, and the Thai government is collaborating with the Japanese government to scale up methane production.

Based on recent studies, Thailand has five potential places (reservoirs) which have the ability to capture carbon emissions. These places are North Malay Basin (1), Kra Basin (2), West Kra Basin (3), Sinphuhorm and Namphong fields (4), and Phitsanulok Basin (S1) (5). The potential CO_2 source and sink locations in Thailand are shown in Figure 7 (MOE, 2022).

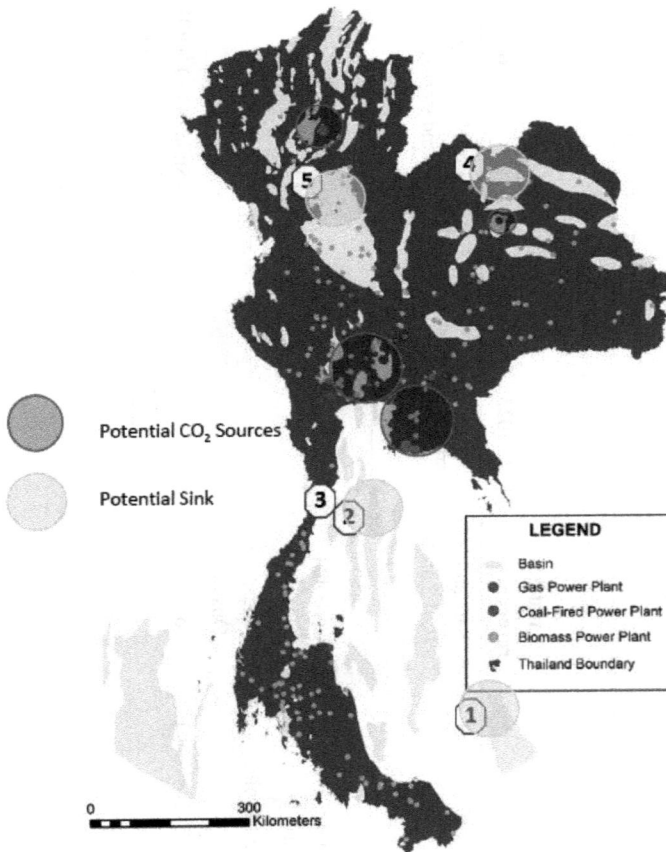

Figure 7.   Potential CO_2 source and sink locations in Thailand.

3.3. *Actions of Thailand Green Growth Roadmap*

Thailand has undertaken various actions and strategies to enhance CCUS development, as follows.

3.3.1. *Storage*

The actions of the carbon storage project are planning and demonstration toward expanded storage capacity, which involves formulating future plans while promoting public acceptance, as well as initiating feasibility studies for demonstration in different locations, as well as coordinating with stakeholders.

3.3.2. *Utilization*

Deciding on the prioritized area is evaluating the utilization potential and relevance of each of the carbon recycled products in terms of CO_2 amount and value of material produced and deciding on prioritized areas.

Concrete and cement in carbon utilization include using public procurement to increase sales and reduce costs, expanding sales channels, establishing new production processes, expanding carbonate use, and establishing carbon-negative concrete and cement. It expands sales by utilizing public procurement and local governments to achieve mass production and economies of scale. It also creates a new product with improved rust prevention performance, broadens its application to building and concrete blocks, considers expanding demand in the private sector to support the introduction of standardization, and gains and expands applications to achieve a circular economy. Furthermore, it encourages existing players, both foreign and domestic, to adopt a CO_2-based concrete and cement production process through investment incentives such as financial support.

Carbon recycled fuel's impact as an alternative aviation fuel is cost reduction and supply expansion via large-scale demonstration. It involves conducting large-scale demonstrations for mass production to achieve cost reductions comparable to existing fuels and increase the supply of competitive sustainable aviation fuel to Thai and international aircraft alike.

Synthetic fuel actions include supporting large-scale synthetic fuel and technological development, introducing government incentives and financial support, deregulating fuel supply agreements to encourage competition and innovation, and developing and inviting foreign carbon recycling players. Manufacturing equipment suitable for large-scale synthetic fuel production will be developed and established to make the same cheaper than gasoline, make existing technologies more efficient to reduce costs, aim for self-sustaining commercialization of synthetic fuel production, encourage R&D of technologies to produce synthetic fuel for practical use, restructure markets through deregulation to attract foreign investors and give local players incentives to establish domestic synthetic fuel production, and offer tax incentives to attract foreign investors and manufacturers to increase market competitiveness and encourage development of innovative synthetic fuel production technologies.

Carbon recycled fuel: Synthetic methane actions include cost reduction and supply expansion through technological development, as well as introducing government incentives and financial support. Doing so will develop technologies to increase the size and efficiency of methanation facilities, increase production capacity to achieve economies of scale, reduce the cost of synthetic methane to that of current LNG prices, inject synthetic methane into existing infrastructure, and convert gas to CN. It will also encourage R&D of manufacturing equipment to produce synthetic methane for practical use.

Implementing demonstration projects for large-scale production of green LPG, which is the carbon recycled fuel's action, will develop basic technologies, such as catalysts that can be commercialized, to implement in demonstration plants for large-scale production of green LPG in Thailand.

The actions of carbon recycling chemicals: Plastic raw materials by artificial photosynthesis include the rapid development of high-conversion-efficiency manufacturing processes and their practical application, as well as tax breaks and investment incentives. It means conducting large-scale demonstrations to allow for higher efficiency production of recycled carbon, reducing costs for the same to equivalent prices as for existing mass-produced plastic, and encouraging domestic players to produce plastic raw materials through tax incentives and investment promotion measures.

The action of carbon recycling chemicals: Bio-manufacturing is the establishment of bio-manufacturing technology. We reduce bio-manufacturing costs using biomass by developing industrial microorganisms using genome editing, etc., as well as developing and demonstrating production processes for commercialization. We develop microbial strains for culture and establish technologies for bio-manufacturing using atmospheric CO_2 as feedstock for practical applications.

The action of the other technologies is the application of DAC technology. We assess the relevance and potential of this and other CCUS technologies, and formulate plans; it also supports the development of new technologies, starts demonstration projects, and provides support for their introduction through cost reduction and subsidies.

4. Potential Challenges to CCUS

While society as a whole knows enough about CCS with a high need for carbon capture, the biggest obstacle to CCUS is cost. As a result, gaining support for the initiative is still a huge issue. The equipment and energy required for the capture and compression processes often have the highest costs. Thus, capturing CO_2 may reduce the efficiency of industrial and power facilities and increase their water use, making such projects ultimately financially unviable. Transferring captured CO_2 presents further difficulties. Pipes are expensive to build and demand much energy for compressing and chilling CO_2 therein.

Financial returns on CCS projects are also riskier than conventional operations because implementation is still in its early phases. As a result, investors apply larger risk premiums, which further raise the private cost of investment. Reducing investor risk is thus essential to encouraging CCS investment and growth. RDD policies that can reduce such risk associated are therefore greatly desired, along with those that can promote innovation, reduce costs, and scale up deployment. Government is also a major impediment to CCUS, however, because of policies and initiatives for reducing emissions.

On the other hand, CCUS has the potential to decarbonize industries (Bosman, 2021; Kawai *et al.*, 2022). It is also cost-effective and expandable in proportion, capable of capturing up to 90 percent of GHGs produced by power plants. Some key challenges in the development of CCUS technology are highlighted in subsequent subsections.

4.1. Challenges and description of Thailand Green Growth Roadmap

This subsection discusses the description of and challenges to the Thailand Green Growth Roadmap considering and focusing on CCUS development in Thailand. The technology-related challenges are as follows:

4.1.1. Carbon storage

The challenges of carbon storage stem from limited research into CO_2 storage capacity. Despite the fact that MoEN has previously conducted studies on the potential of CCS, insights gained from pilot projects and technology demonstrations are still limited.

4.1.2. Carbon utilization

Large-scale utilization of captured CO_2 for producing carbon recycling material has not commenced in Thailand, and the potential and relevance of each of the products are still unclear.

Challenges to the use of captured CO_2 in cement and concrete are establishment of production technology, product development, and cost reduction. Despite being one of the leading cement and concrete procedures in the world, there has not been much movement regarding the utilization of CO_2 in technologies for cement and concrete production. Also, green initiatives in this sector have focused on developing recycled plastic for roads and hydraulic cement for construction.

For carbon-based alternative aviation fuel, challenges again are establishing the production technology and reducing costs. Currently, biojet fuel is imported to Thailand from overseas. The development of biofuel for the aviation sector has been an ongoing initiative through collaboration between big players in Thailand, PTT and Aerothai, and a foreign enterprise, Boeing. However, due to the high cost of production, there has been a challenge in establishing a production technology that will be suitable for commercialization and mass production. High production costs have also translated into higher prices. Therefore, further reducing adoption costs to be on par with existing jet fuel is crucial.

The establishment of production technology, cost reduction of synthetic fuel, and capacity development are the challenges of producing synthetic fuel from carbon-recycled fuel. Due to the regulated nature of the fuel market and rigid supply agreements in Thailand, the development of production technologies and process gases has been slow. One of the big projects existing in Thailand is R&D into the use of non-edible feedstock for synthetic fuel production. Production process costs must be reduced, however, to encourage more players to get involved and increase adoption.

The challenges to synthetic methane are technological development and limited government and financial support. Synthetic methane has stagnated in Thailand due to a lack of recognition of its benefits.

Establishing production technologies is the challenge of green LPG. There is a need to develop production technologies for green LPG in Thailand, as it imports much of this fuel, driving costs higher. Bio LPG development has also been slow due to limited government support and financial incentives for investment.

Challenges to carbon recycling chemicals are technological development, cost reductions, and developing domestic OEMs. Although roadmaps have progressed to the development stage, commercialization remains a challenge due to financial constraints. Most of the technologies as well as the specialty chemical businesses in Thailand are imported from overseas or owned by foreign companies. Due to a lack of technology and knowledge in the area, there has been a lack of domestic players in the sector, which is coupled with a lack of incentives for investment.

Cost reduction and commercialization are the challenges of carbon recycling chemicals in bio-manufacturing. The main production of biofuel, i.e., bioethanol, has been focused on first-generation production using palm oil as feedstock, which is abundant in Thailand. The Thailand Integrated Energy Blueprint (TIEB) 2015 sets the production target for second and third-generation biofuels at 10 ktoe by 2036. The government supports this objective by supporting research at Thai universities. In addition, to reduce production costs, PTT Think Algae and TISTR conducted integration programs into oil production from microalgae using co-processing with wastes such as water and flue gas. The crude algal oil obtained from the TISTR pilot plant has been continuously sent to PTT Research and Technology Institute (PTT RTI) for developing proper formulations of biofuels for commercialization. The challenge is the high cost of biomass compared to first-generation biofuels.

As new goals for achieving 2050 carbon neutrality and 2065 net-zero emissions have been defined, carbon reduction initiatives must be accelerated in addition to existing initiatives.

5. Conclusion

Carbon capture usage and storage (CCUS) technology is critical in addressing rapidly expanding energy demand and resulting carbon emissions increases, which in turn drive climate change. This chapter has provided an overview of CCUS, its worldwide policy implications, and its specific implications for Thailand, with its four sections organized as follows. First is an overview of CCUS and the carbon capture, utilization, and storage chain. Second is overall global policy implications, showing the status of CCUS in different countries and highlighting the progress. Third is an examination of Thailand's CCUS landscape, describing the Thailand Nationally Determined Contribution Plan for CCUS technology development in recent years. The chapter also showcases pilot CCUS projects in Thailand, including oil and gas reservoirs, timelines, and expected outcomes. Five pilot projects are being studied: the Arthit, Kra Basin, Phu Horm, S1, and BLCP Pilot Projects. Lastly, this chapter explained the potential barriers, descriptions, and challenges for CCUS under the Thailand Green Growth Roadmap.

While Thailand has implemented various mitigation options across sectors, including electric vehicles, biofuels, green building practices, conservation, and renewables, there is an emerging consensus that CCUS is an important decarbonization technology. CCUS development and commercialization are necessary to drive further enhancements and cost reductions, with the government playing a crucial role in creating sustainable markets through supportive policies.

It is evident that markets alone cannot make CCUS a clean energy success story. Government and industry must seize the opportunity to work together to achieve energy and climate goals, recognizing the potential environmental and economic benefits that CCUS offers.

References

Bajpai, S., Shreyash, N., Singh, S., Memon, A. R., Sonker, M., Kr Tiwary, S., and Biswas, S. (2022). Opportunities, challenges and the way ahead for carbon capture, utilization and sequestration (CCUS) by the hydrocarbon industry: Towards a sustainable future. *Energy Reports, 8*, 15595–15616. https://doi.org/10.1016/j.egyr.2022.11.023.

Bhatia, S. K., Bhatia, R. K., Jeon, J. M., Kumar, G., and Yang, Y. H. (2019). Carbon dioxide capture and bioenergy production using biological system — A review. *Renewable and Sustainable Energy Reviews, 110*, 143–158. https://doi.org/10.1016/J.RSER.2019.04.070.

Bosman, T. (2021). *The Potential of Carbon Capture, Utilization and Storage (CCUS) to Decarbonize the Industry Sector in Europe.* Oxford: The Oxford Institute for Energy Studies.

IEA (International Energy Agency). (2020). CCUS in the transition to net-zero emissions — CCUS in Clean Energy Transitions — Analysis — IEA. https://www.iea.org/reports/ccus-in-clean-energy-transitions/ccus-in-the-transition-to-net-zero-emissions.

Kawai, E., Ozawa, A., and Leibowicz, B. D. (2022). Role of carbon capture and utilization (CCU) for decarbonization of industrial sector: A case study of Japan. *Applied Energy, 328*, 120183. https://doi.org/10.1016/J.APENERGY.2022.120183.

Lambert, T. H., Hoadley, A. F., and Hooper, B. (2016). Flexible operation and economic incentives to reduce the cost of CO_2 capture. *International Journal of Greenhouse Gas Control, 48*, 321–326. https://doi.org/10.1016/J.IJGGC.2016.01.023.

Li, Q., Song, R., Liu, X., Liu, G., and Sun, Y. (2016). Monitoring of carbon dioxide geological utilization and storage in China: A review. *Acid Gas Extraction for Disposal and Related Topics, 225*, 331–358. https://doi.org/10.1002/9781118938652.CH22.

Li, Q., Wei, Y. N., and Chen, Z. A. (2015). Water-CCUS nexus: Challenges and opportunities of China's coal chemical industry. *Clean Technologies and Environmental Policy, 18*(3), 775–786. https://doi.org/10.1007/S10098-015-1049-Z.

MOE (Ministry of Energy). (2022). Work Progress on CCUS Development Report.

UN Global Compact. (2020). Thailand's Long-term Greenhouse Gas Emission Development Strategy, presented at COP26, Thailand Pavilion. https://globalcompact-th.com/news/detail/602.

Chapter 4

On the Role of Startups in Global CCUS Development and Deployment: An Exploratory Analysis from the Southeast Asian Perspective

Wasim Ahmad[*,‡]**, Manmeet Kaur**[*,§]**,**
Nazneen Mansoori[*,¶]**, and Han Phoumin**[†,ǁ]

*Department of Economic Sciences, Indian
Institute of Technology Kanpur, Kanpur, India*

†*ERIA (Economic Research Institute for ASEAN
and East Asia), Senayan, Jakarta Pusat, Indonesia*

‡*wasimad@iitk.ac.in*

§*manmeetk@iitk.ac.in*

¶*nazneenm@iitk.ac.in*

ǁ*han.phoumin@eria.org*

Abstract

Carbon emissions concerns have made carbon capture, utilization, and storage (CCUS) an alternative to utilizing carbon for industrial and non-industrial purposes. Efforts have been made despite limitations to understand the CCUS ecosystem in the Association of Southeast Asian Nations (ASEAN). The main aim of this study is to understand the CCUS ecosystem globally and draw lessons for ASEAN. It begins by covering existing CCS frameworks in different countries and drawing learnings for ASEAN. Then, it focuses on the exploratory role of cleantech start-ups, particularly firms directly involved in CCS development and deployment. Funding, patents, and trademarks complement each other, and the role of human capital, i.e., education and founders' experience, is visible, making the role of educational institutions crucial. It is advisable to devise provisions for knowledge collaboration and inter-university accelerator programs to promote entrepreneurship and develop programs to train the necessary human resources. The study cites the need to promote professional expertise, for which ASEAN may create a separate fund for developing and deploying CCUS through endogenously grown startups.

Keywords: Carbon capture, utilization, and storage (CCUS), sustainable development, startups, negative binomial regression, energy regulations.

1. Introduction

The Paris Agreement and the United Nations Framework Convention on Climate Change's (UNFCCC) Conference of the Parties (COP 26) have reignited the debate over climate mitigation strategies. Carbon neutrality has also forced countries to act immediately against climate concerns. Carbon capture, utilization, and storage (CCUS or CCS, used interchangeably) is gaining attention due to widespread application. Its adoption and expansion have many benefits in terms of climate and livelihood risk. Widespread application of carbon-based products has also attracted researchers interested in understanding and analyzing the growing need for CCUS. Economically, CCUS helps reduce carbon emissions from many environmentally sensitive industrial activities, including steel, cement, iron, chemical, power generation, agriculture, and transportation. Carbon capture from these industries safeguards the environment and ensures sustainable manufacturing and livelihoods.

According to Turan *et al.* (2021), widespread CCUS adoption as an effective measure of carbon emission reduction is still a major challenge as several countries have not yet formalized or adopted climate change mitigation. Renewable energy still has a larger share than CCUS. Article 6 of the Paris Agreement guides voluntary CCUS adoption in carbon emissions reduction.

On the bright side, carbon capture allows storing and transporting captured carbon for further usage, such as building materials and consumer goods, often called valorization. With the widespread usage of carbon as raw materials, several countries have started investing in CCUS facilities, but the list is still small. The leading countries are the US, UAE, South Africa, Norway, Canada, China, Egypt, Iran, Iraq, Malawi, Saudi Arabia, and Bahrain. COP 26 is driving further CCUS adoption.

Our approach in this study is to answer the following questions regarding CCUS development and deployment in Southeast Asia and other emerging economies. First, what is the organic composition of CCUS-linked startups globally and in Southeast Asia? In other words, how are innovations of startups helping global CCUS development and deployment? Second, which factors drive the CCUS startup ecosystem in areas other than Southeast Asia? Third, what are the key lessons for ASEAN?

We aim to investigate carbon risk reduction initiatives and how far Southeast Asian economies have planned to encourage CCUS adoption. In this context, we aim to highlight the best CCUS adoption practices by startups globally and lessons for Southeast Asia. These questions are linked to developing and deploying new CCUS technologies and how the startup ecosystem helps CCUS adoption expand through its innovation and financing sources. Our startup analysis may also provide new directions for channeling investments and help design policy guidelines to promote CCUS-based practices and lessons.

Home to 630 million people, Southeast Asia comprises 11 small and medium-sized individual economies. ASEAN is a significant regional bloc with a combined market value of USD 2.4 trillion (2015). While it is often compared with the Eurozone due to its socioeconomic environment, ASEAN member states are financially independent and maintain their own currencies. Socioeconomic complementarity is also one of the biggest strengths of ASEAN. Its leading economies include Singapore, Indonesia, Malaysia, Thailand, Vietnam, and the Philippines. The region is focusing on environmental sustainability and economic growth. Southeast Asia is also home to leading companies in the manufacturing and services sectors with a sizable number of startups. Strong economic integration among member states provides scope for understanding preparedness for global challenges, including climate risk. As some member states are vulnerable to climate risk, it is essential to emphasize the credible role of new developments, including CCUS deployment, which, though at a nascent stage, may augment climate risk mitigation efforts by adopting best practices.

In this study, we add the following dimensions to the existing wisdom on CCUS. First is the status of current CCUS developments and analysis of best practices by accounting for the strengths, weaknesses, and lessons learned. Second is the financing of CCUS adoption and deployment. Finally, we add the dimension of CCUS development by examining startup ecosystems in major economies. We also provide sufficient details about the organic composition of CCUS startups and prepare a roadmap for Southeast Asia. Some startups in the US, China, and Australia are relevant for CCUS development and deployment.

2. Background and Related Literature

As mentioned above, CCUS is gradually becoming an indispensable carbon mitigation strategy due to the win–win situation it offers. Widespread use of carbon products in households and industry is attracting attention. Rising climate change and carbon reduction concerns have encouraged countries to take credible measures. The European Union (EU) and some 45 nations have agreed on credible zero-emission targets (EIA, 2021). Some nations have issued guidelines and formulated policies. Multilateral platforms such as The Climate Ambition Alliance bring together countries to target zero net emissions by 2050.

CCUS, in this regard, is critical to meet zero emissions targets as it helps reduce emissions arising from core industries such as coal, steel, cement, and other heavy industries for deep underground storage and reuse. Figure 1 gives an overview of the global status of CCS. Most projects are in either early or advanced development stages, reflecting the

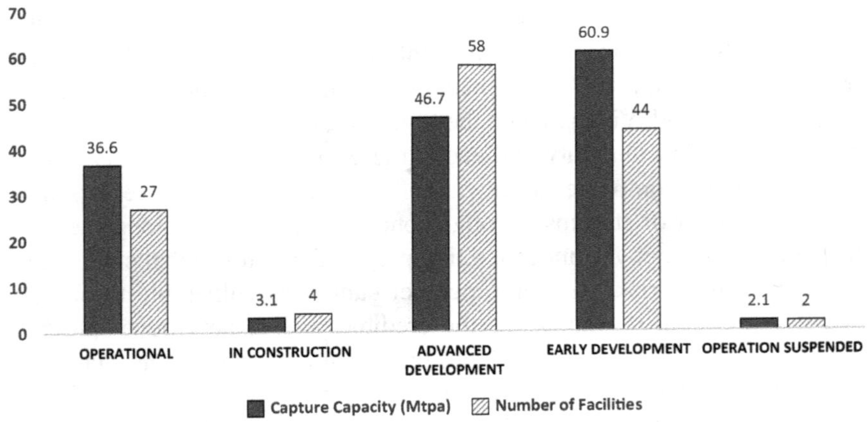

Figure 1.   Global CCS status.

Source: Global Status of CCS 2021, Global CCS Institute.

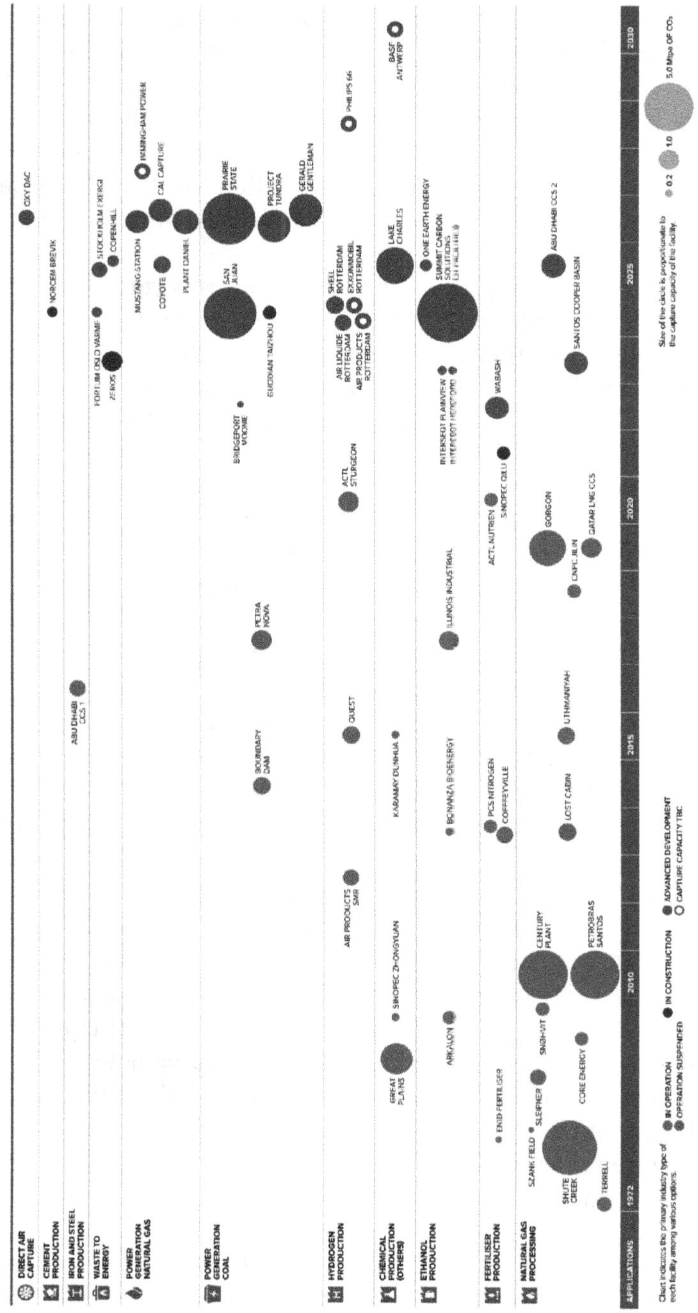

Figure 2. Sector-wise CCUS project deployment.

Source: Global CCS Institute.

growing focus on CCUS deployment. Figure 2 shows how CCUS projects can be deployed, including direct air capture, cement and chemical production, power generation, fertilizer, and natural gas. As all of these industries emit carbon, carbon capture may help immensely.

Some CCUS projects have greater capacities than others. For example, Shell's Rotterdam hydrogen project has millions of tons. For smaller ASEAN economies, Norway is a more appropriate example. The Norcem Brevik project is linked to a cement manufacturing unit, which may apply to ASEAN. It is expected to provide major learning opportunities for other economies. Anderson and Newell (2004) estimate deploying carbon storage from flue gases and find it to cost USD 200–250 per ton of carbon. Liu and Gallagher (2010) provide a detailed account of China's low-carbon economy and a CCS deployment roadmap. They recommend streamlining investment in CCS through development, demonstration, adoption, and effective integration. Fennell et al. (2012) provide a literature summary of CCS in cement, iron and steel, refinery, and biomass, finding that CCS costs vary from USD 10–100 per ton of carbon. Lilliestam et al. (2012) compare CCS-linked thermal power plants with Concentrating Solar Power (CSP) and recommend favoring CSP by highlighting the difficulties both technologies face. Liang and Li (2012) provide recommendations for decarbonizing China's cement industry, recommending a CCS-ready scenario for new cement plants using Guangdong province data. Li et al. (2013) analyze the successes and limitations of CCS deployment in the cement sector, finding that the most efficient cost is USD 60/ton of carbon, and citing the need for public funds to support such initiatives.

Rootzén and Johnsson (2015) examine core steel, cement, and petroleum refining sectors in the Nordic countries of Finland, Sweden, Denmark, and Norway, finding that deploying CCS would significantly reduce carbon emissions and might also help meet climate change targets. Li et al. (2017) use the dynamic computable general equilibrium (CGE) model to examine whether adopting electric vehicles (EV) and CCS will help meet carbon reduction targets, finding that doing so may significantly minimize climate risk. Leeson et al. (2017) provide a detailed literature review of CCS in iron, steel, cement, and oil refining, finding that the principal cost was associated with CCS deployment and operation.

Bui et al. (2018) deal with some nagging issues relating to limited deployment of CCS despite its benefits, finding that the dependence on the private sector for integration and maintenance across capture, storage, and transportation chains is a formidable reason. They also highlight reasons why there should be public support for CCS, including cross-chain default, carbon storage and performance risks, carbon storage permit, and insurance markets. In this context, the role of the public sector can be encouraged through startup promotion as more new firms innovate and scale. The government can also promote these efforts through carbon emission reduction initiatives. Hills *et al.* (2019) provide a detailed account of CCS applicability to the cement sector and highlight the challenges that may make it unsuitable for CCS. Guo and Huang (2020) propose a new CCS project investment model that may help meet the 2050 climate change target and simultaneously provide a sustainable solution to the carbon reduction challenge, suggesting that CCS investment should get priority by 2035 by avoiding early large-scale demonstrations.

Kearns *et al.* (2021) focus on economies of scale, energy cost, and technological innovation, providing a detailed account of the CCS value chain and factors determining capture, compression, transport, and storage costs. A lack of focus on CCS innovation and startup ecosystem is a serious limitation, however. Paltsev *et al.* (2021) examine CCS deployment in cement, iron, steel, and chemicals, applying the MIT Economic Projection and Policy Analysis (EPPA) model to take the temperature of projects both with and without CCS and finding that non-CCS scenarios increase temperatures. Their analysis provides a suitable roadmap for enhancing CCS employment in these critical sectors. Poudyal and Adhikari (2021) suggest how carbon emissions in the cement industry can be captured and reused as non-calcium carbonate via nanotechnology. Stokke *et al.* (2022) examine a Norwegian cement producer firm that has developed low-carbon cement with CCS, finding that early procurement suppliers have a direct impact on green public procurement. They suggest reducing climate change globally via this approach.

While CCS development and deployment face challenges, the technology has merit in reducing carbon emissions. We find that over-dependence on the private sector and lack of public sector support have significantly affected CCS deployment. Economies of scale and energy consumption are other critical challenges in sustainable development. This broad overview of the literature suggests a need to study the CCS ecosystem through recent developments, especially how new firms adapt to the challenge of CCS adoption and which sectors they focus on. This broad account of the diversification of CCS deployment and innovation-based solutions may further enhance the understanding of climate risk mitigation and how Southeast Asia can reap the benefits. Literature on Southeast Asia is still evolving, and some studies provide useful recommendations worth revisiting. Phoumin *et al.* (2020) focus on green hydrogen and how CCS can make it economical, providing a deep understanding through energy modeling. Nepal *et al.* (2021) provide a detailed overview of how CCS could be useful in ASEAN and help the region to continue burning fossil fuels and hydrogen as green energy. Another similar study by Chang and Phoumin (2022) highlights the role of CCS and the utility of blue hydrogen.

Most of these studies are country- and project-specific, and none have examined the role of startups in augmenting CCS. As climate transition is a long-term challenge, the present study may further the CCS literature, providing a diverse dimension to CCS development and deployment, while also contributing to the cleantech literature.

3. Data and Methodology

3.1. *Design*

The primary aim of this study is to examine and analyze best practices in CCUS development and deployment through cleantech startups. The study is divided into two stages. The first stage encompasses a global perspective and comparative analysis of different policy measures and how they can help Southeast Asia design or align its CCS development and deployment efforts. In this context, the study analyzes significant global CCS projects and existing projects in Southeast Asia and provides new directions and an up-to-date understanding of CCUS development and deployment frameworks. It uses policy documents sourced through different sources and explores the resources of the International Energy Agency (IEA) and Global CCS Institute. Additional information was collected by scraping other web sources, the procedure for which will be elaborated on hereinafter.

In the second stage, the analysis scrutinizes the exploratory role of cleantech startups, particularly firms directly involved in CCS development and deployment. This is critical because of the potential for Southeast Asian economies with CCS and their financing channels. There is a need to expand the scope of CCS expansion by focusing on the potential of startups in CCS deployment. In this study, 51 startups and firms located in the US, the UK, Canada, The Netherlands, Norway, Israel, Finland, Denmark, Switzerland, and China are considered, with data sourced from Crunchbase. These startups and firms are filtered based on their activities in the CCS domain. Table 1 provides the list of firms considered for analysis. Notably, there is a high concentration of firms in the US, followed by the UK and Canada, highlighting the dominance of the US economy in CCS-linked startups. The main task is to analyze the spread and activities of firms in CCS. The study examines the factors determining these firms' CCS development and deployment. The number of patents acquired by these firms and the domains in which firms seek further intellectual support are critical components of innovation and support for CCS. Lessons from these startups, especially the critical factors determining their number of patents, are key to creating a knowledge economy. Our approach uses hand-picked data on different factors, with relevant data on firms and their promoters being assimilated. The variables and their sources are as follows:

Patents and Trademarks: Total number of patents and trademarks registered for a given firm. *Source*: Crunchbase.

Funding: Natural log for total funding received by firms (in USD). *Source*: Crunchbase and authors' calculations.

Country readiness score: An index developed by the Global CCS Institute based on storage readiness, laws and regulations, and CCS policy. *Source*: CCS Readiness Index Report.

Age: Time in months since firms' founding as of this writing. *Source*: Crunchbase and authors' calculations.

Average PIS: Premier Institute Score (PIS) is a score for the institute from which promoters received their highest qualifications. Scores are in descending order by ranking, ranging from 10 for Top-100 ranked institutes to zero for institutes not recorded in 2021 QS rankings. The PIS for each firm is then computed by averaging the total firm score by the number of founders. *Source*: Manual data collected from founders' LinkedIn page and authors' calculations.

Table 1. List of firms.

Name	Country	Industries
Carbon Direct	United States	Environmental Consulting, Environmental Engineering, Sustainability
Climeworks	Switzerland	CleanTech, Environmental Engineering, GreenTech
Group14 Technologies	United States	Battery, Electric Vehicle, Energy Storage
Wildcat Discovery Technologies	United States	Clean Energy, CleanTech, Energy, Product Design
Carbon Clean Solutions	United Kingdom	CleanTech, Oil and Gas, Pollution Control
Svante	Canada	Clean Energy, CleanTech, Environmental Engineering
Pond Technologies	Canada	Animal Feed, CleanTech, Environmental Consulting, GreenTech
Summit Carbon Solutions	United States	Fossil Fuels
Brilliant Planet	United Kingdom	CleanTech
Heimdal	United Kingdom	Chemical, Renewable Energy, Water
Carbon Capture	United States	Environmental Consulting, Renewable Energy
Carbon America	United States	Environmental Engineering, Renewable Energy, Sustainability
NAWA Technologies	France	CleanTech, Energy, Nanotechnology
CleanO2	Canada	Business Development, Energy, Energy Efficiency, Technology
Holy Grail	United States	Energy, Energy Efficiency, Energy Storage, Renewable Energy
44.01	United Kingdom	Biofuel, Environmental Engineering
Chasm Technologies	United States	Consulting, CRM, Information Technology, Manufacturing, Robotics
Skyonic	United States	Clean Energy, Energy, Environmental Engineering, Renewable Energy
Seabound	United Kingdom	Clean Energy, Environmental Consulting, Environmental Engineering
Elysian Carbon Management	United States	Facility Management, Industrial, Manufacturing, Project Management
Hydrogen Mem-Tech	Norway	Renewable Energy
Algiecel	Denmark	Biotechnology, GreenTech, Pollution Control, Sustainability
Skytree	The Netherlands	AgTech, Air Transportation, GreenTech, Sustainability
RepAir Carbon Capture	United States	Environmental Consulting, Renewable Energy

Reverion	Germany	Energy
Carbon Recycling International	Iceland	Energy, Energy Efficiency, Energy Storage, Fuel
CollectiveCrunch	Finland	Analytics, Artificial Intelligence, Energy Efficiency
Aqualung Carbon Capture	Norway	Industrial, Renewable Energy
C2CNT	Canada	Energy Efficiency, Energy Management, Energy Storage
Humble Midstream	United States	Energy, Oil and Gas, Renewable Energy
Oakbio	United States	Biotechnology, GreenTech, Product Research
Captura	United States	Environmental Consulting
Tonik Energy	United Kingdom	Energy, Energy Efficiency, Renewable Energy
EnerG2	United States	Advanced Materials, Chemical Engineering, Energy, Energy Storage
Surface Transforms Plc	United Kingdom	Automotive, Manufacturing
Gravitas Infinitum	United States	Renewable Energy Semiconductor Manufacturing
Saratoga Energy	United States	Automotive, Chemical, Electronics, Energy, Manufacturing
Airovation Technologies	Israel	Clean Energy, CleanTech, GreenTech, Smart Home
PETROLERN	United States	Clean Energy, Energy, Oil and Gas
MOF Technologies	United Kingdom	Chemical Engineering, Clean Energy, Nanotechnology
UP Catalyst	Estonia	Nanotechnology
Saint Jean Carbon	Canada	Energy, Renewable Energy
Hyperion Global Energy	Canada	Energy, Industrial, Recycling
HempConnect GmbH	Germany	Agriculture, Farming, PaaS, Sustainability
CarbonPoint Solutions	United States	Energy Efficiency, Environmental Consulting, Information Technology
Vizopay	United Kingdom	Sustainability
Legend Energy	China	CleanTech, Energy Storage
Genvision	United States	Artificial Intelligence, CleanTech, GreenTech
Ziknes	Spain	Machinery Manufacturing, Manufacturing, Robotics
Power-to-Gas Hungary Ltd.	United States	Energy, Environmental Consulting, Renewable Energy
Agricarbon	United Kingdom	Agriculture, Farming, Sustainability

Table 2. Summary statistics.

Variables	N	Mean	Std. Dev.	Min	Max
Funding	51	15.746	2.472	11.301	20.480
Average PIS	51	6.776	3.700	0	10
Average industry experience	51	10.747	7.708	0	31
Country readiness score	50	63.780	14.088	19	72
Specialization score	51	1.078	1.074	0	5
Prior startup experience	51	0.490	0.505	0	1
Founders (in numbers)	51	1.725	0.918	1	5
Age	51	137.745	212.271	12	1476
Patents and trademarks	51	5.451	15.468	0	97

Average industry experience: Total industry experience in years of founders for each startup averaged by the number of founders. *Source*: Manual data collected from founders' LinkedIn pages and authors' calculations.

Specialization score: Sum of founders having an educational background in a field relevant to CCUS, such as chemistry, physics, biochemistry, or other such science or engineering-related interdisciplinary education. *Source*: Manual data collected from founders' LinkedIn pages and authors' calculations.

Prior startup experience: Dummy variable, set to 1 for startups having at least one promoter with prior experience as a startup founder. *Source*: Manual data collected from founders' LinkedIn pages.

Founders: The total number of founders of each firm, with data discrepancies from the database addressed. *Source*: Crunchbase and web scraping.

Summary statistics for the variables are given in Table 2. The oldest firm in the sample was established in 1990, giving the maximum number of months since founding. The country readiness score has only 50 observations as no index has been reported for Israel. The highest number of total patents and trademarks registered for a firm is 97, while the mean is close to five. Some 49% of firms have at least one founder with prior founding experience, while firms have approximately two founders on average.

3.2. *Preliminary analysis*

For Southeast Asia, Kimura *et al.* (2021a) highlight three critical dimensions for efficient deployment. Figure 3 highlights issues related to CCS development in ASEAN. Foremost is knowledge sharing, which is directly linked to CCS development. CCS deployment requires technological collaboration, better financing, and inter-country technology transfer. Effective regulation encourages innovation. Risk management and practical experiences are linked to CCS deployment. These factors are dependent on regulatory frameworks. Figure 4 complements Figure 3, highlighting the regulatory dimensions required for CCS deployment in ASEAN (Kimura *et al.*, 2021b). However, one major limitation of these studies is that they ignore the role of cross-country collaboration and the potential for innovation by startups. Endogenous growth capabilities must

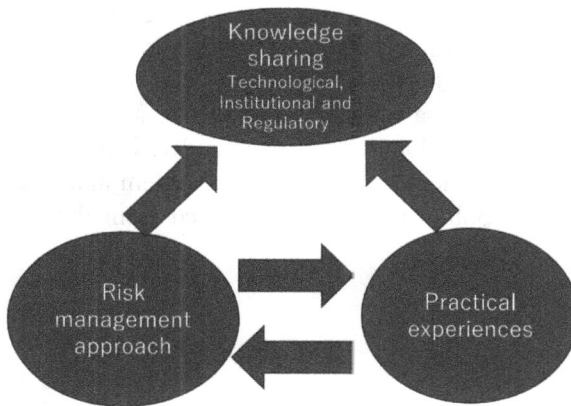

Figure 3.   Issues with ASEAN CCS deployment.
Source: Kimura, S., Shinchi, K., Kawagishi, S., & Coulmas, U. (2021a), page # 36.

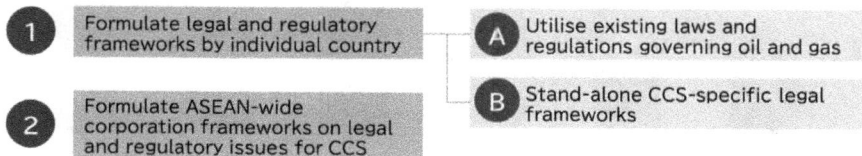

Figure 4.   Regulatory dimensions of ASEAN CCS deployment.
Source: Kimura *et al.* (2021b), p. 30.

be promoted to enhance climate change mitigation. By contrast, the main aim of this study is to highlight areas in which cleantech startups can help expand CCS development and deployment horizons.

Progress of Southeastern CCS projects shows Indonesian and Malaysian interests. The 2.5 Mtpa project in Sakekamang, South Sumatra, Indonesia, is expected to capture carbon from Repsol's natural gas processing plant. At the same time, PAU Central Sulawesi Clean Fuel Ammonia Production with CCUS is another significant Indonesian project. The Petronas Kasawari Gas Field Development Project in Malaysia is under development. Japan and China are two major destinations in the Southeast Asia region with enormous lessons and technology spillover to offer.

Trends appear to favor developed nations more than developing ones, as observed from the sampled firms. Figure 4 gives a roadmap to focus on CCS development and promote innovation in Southeast Asia. As the net-zero debate intensifies, the race for better climate risk management can only be won by Southeast Asian economies when they also incorporate such developments in climate change efforts. A low-carbon economy can also be enhanced through these channels.

The prior literature analysis highlights pre-existing trends in the global CCS industry. Further exploration has been undertaken over the study sample to gain insight into the prevailing CCS ecosystem that would strengthen the resolutions and policy implications derived from the study. Figure 5

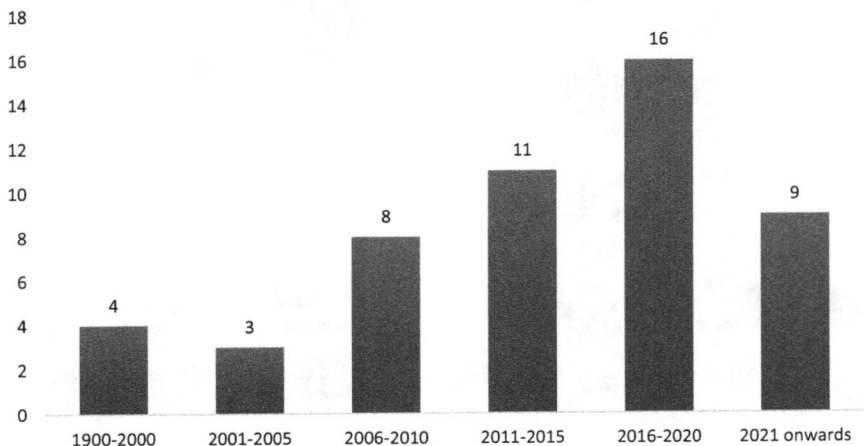

Figure 5.   Frequency of firms by year founded.

Source: Crunchbase, authors' calculations.

depicts trends observable in the data concerning firms' founding dates, with the most firms founded between 2016 and 2020, right after the signing of the Paris Agreement and concurrent with growing climate concerns. Notably, nine firms were founded in 2021 and 2022 alone, highlighting the shift in global attention toward carbon neutrality. Establishing firms in the CCS space requires infrastructure capacity and favorable policy. In this light, the Global CCS Institute developed CCS scores to measure a country's readiness to cater to the needs of the industry. Figure 6 reports readiness scores spatially for countries where sampled firms are located, except for Israel, for which the CCS report doesn't provide a score. The US has the highest score, at 72, closely followed by Canada. Countries with the highest scores are shaded in darker shades of blue. Switzerland has the lowest score, at 19.

Funding received by firms is another critical factor affecting their survival, growth, and success in tandem with the domestic business environment. Figure 7 shows the total funding received by firms by comparing the share of equity funding in total funding. The line represents the log of equity funding, while the bars give the log of total funding. Only seven firms had no equity-based funding, while the rest received most of their

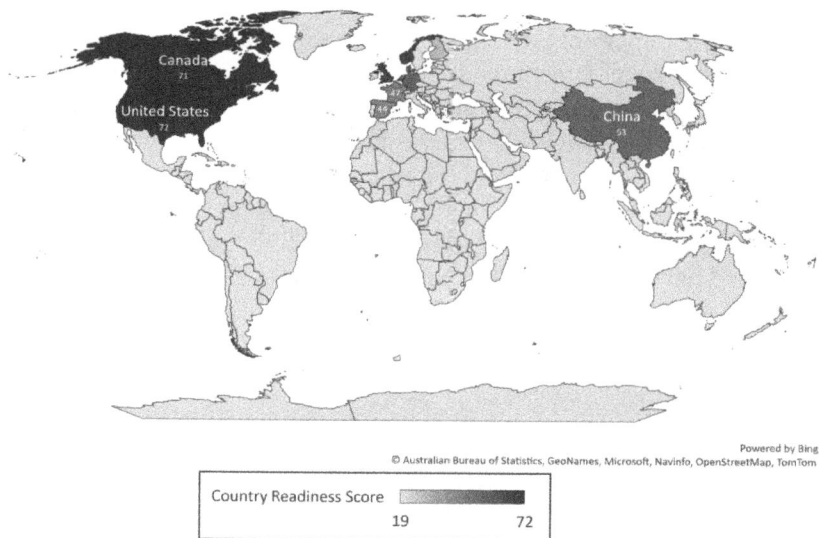

Figure 6. CCUS deployment readiness score by country.
Source: Global CCS Institute.

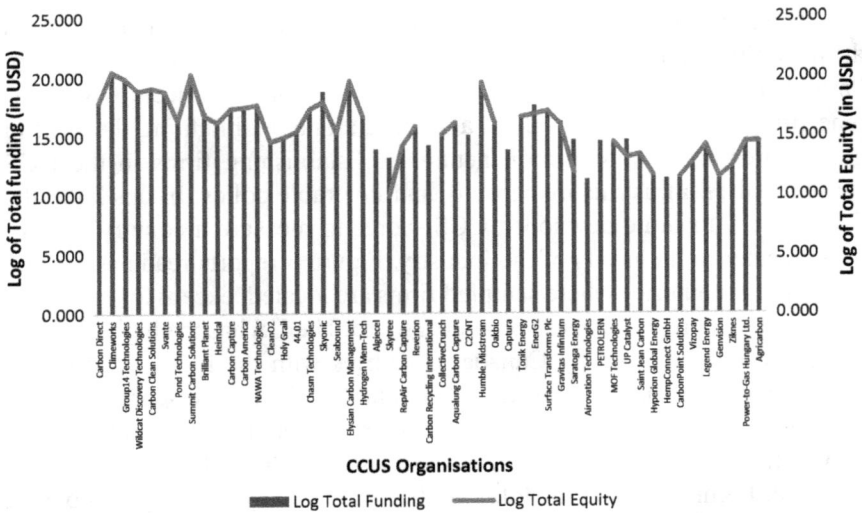

Figure 7.   Share of equity in total funding acquired by firms.

Source: Crunchbase and authors' calculations.

funding as equity funding, demonstrating the global inclination toward equity-based funding for CCS firms.

Another important element is patents and trademarks owned, indicating research, development, and innovations. Table 3 shows innovation in CCS-linked startups as the number of patent and trademark registrations acquired. The trend shows how investments guide innovations in CCS and how firms are trying to expand CCS networks. The United States is home to the firm with the most registered CCS patents, Wildcat Discovery Technologies, followed by Climeworks of Switzerland, which has, 39 registered patents. While patents and trademarks are anticipated to have a signaling effect of potential for building novel technologies, developing and implementing modern-day solutions to global problems require appropriate and timely allocation of resources, financial and physical alike.

Accordingly, we plot the total patents and trademarks acquired along with the log of total funding, showing a definitive relationship between patents, indicating a firm's excellence, and funding, one of the anticipated driving forces for a firm's performance. Figure 8 provides the requisite graphs. The funding variable is observed to move in parallel with the number of patents and trademarks acquired, except for outliers with higher numbers of patents and trademarks.

Table 3.   Number of patents and trademarks registered by firm.

Firm name	Country	Patents granted	Trademarks registered
Climeworks	Switzerland	39	4
Group14 Technologies	United States	17	5
Wildcat Discovery Technologies	United States	92	5
Carbon Clean Solutions	United Kingdom	10	13
Svante	Canada	25	1
Pond Technologies	Canada	6	0
Brilliant Planet	United Kingdom	0	0
Carbon Capture	United States	0	0
Carbon America	United States	0	2
NAWA Technologies	France	11	3
CleanO2	Canada	1	2
Holy Grail	United States	0	0
Chasm Technologies	United States	2	0
Skyonic	United States	8	4
Hydrogen Mem-Tech	Norway	3	0
Algiecel	Denmark	0	1
Skytree	The Netherlands	5	0
CollectiveCrunch	Finland	1	0
Surface Transforms Plc	United Kingdom	6	1
Airovation Technologies	Israel	1	0
MOF Technologies	United Kingdom	4	0
Hyperion Global Energy	Canada	0	0
CarbonPoint Solutions	United States	6	0

Source: Crunchbase and authors' calculations.

Figure 9 shows the proportions of registered patents in various domains. Over half of the patents are registered under basic electric elements, followed by physical/chemical processes of general apparatus. Other classes account for very low proportions, with the lowest in computing and calculating, pointing toward the limited role of computing in CCUS, while the aforementioned electronic and process-related majority

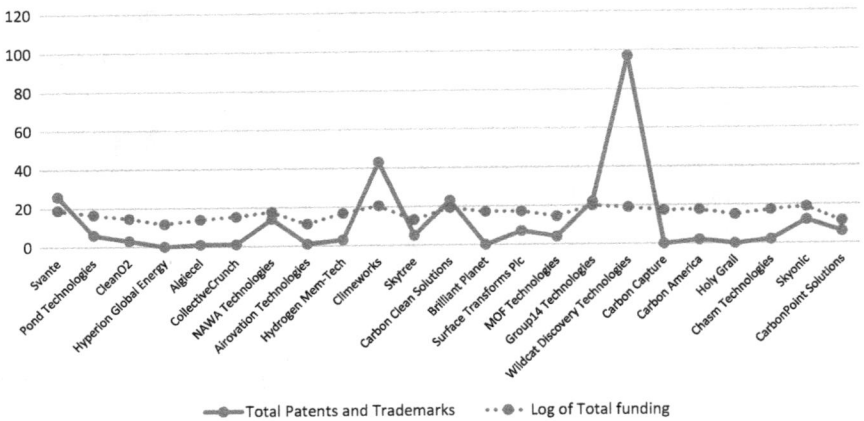

Figure 8.   Total funding of firms with patents and trademarks.

Source: Crunchbase and authors' calculations.

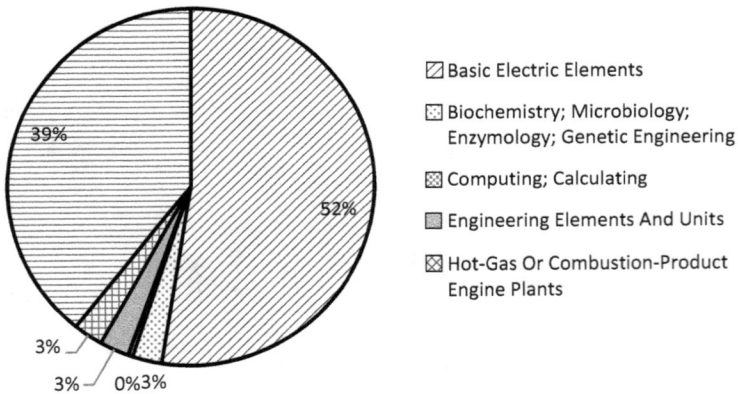

Figure 9.   Classification of acquired patent groups in sample data.

Source: Crunchbase and authors' calculations.

implies the enhanced role of those categories. Policymaking must focus on engaging and augmenting research in these disciplines in CCS.

Further exploration of country data is required for an in-depth investigation of global trends. Figure 10 reports the country-level distribution of patent classes from the sample. The US has the most overall patents,

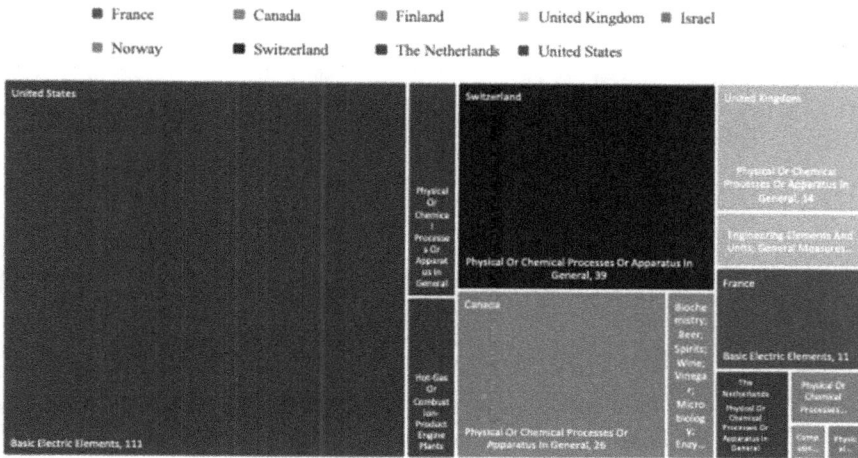

Figure 10. Distribution of patent groups among countries of sample firms.
Source: Crunchbase and authors' calculations.

particularly in the basic electric elements group. Switzerland has the high-est number of patents in physical or chemical processes or general appa-ratus. Also, it has no presence in other patent classes. Next in the physical or chemical processes class is Canada, which also has the highest number of biochemistry patents. The UK shares patents in only two classes, while the remaining countries have patents in only some single patent groups.

Distribution of trademarks by country is given in Figure 11, with the most overall in chemical substances, followed by scientific and techno-logical services. The US has the largest share in chemical trademarks, followed by Switzerland. UK firms primarily have trademarks relating to scientific and technological services. Canada, France, and Denmark are the only remaining countries with relevant trademarks. In contrast, all countries in the sample had some relevant patents.

As the prospects of any business depend on the competence of its founders, the study next turns to this aspect of our sampled firms. Figure 12 groups companies by number of founders. Only one firm has five founders, while 24 firms have a single founder.

Next, the PIS for firms, which indicates founders' education at global elite institutes, is plotted along with funding received.

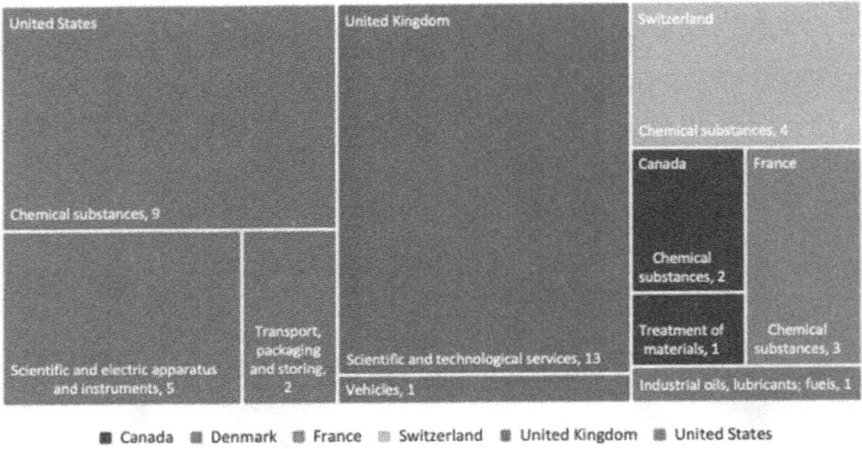

Figure 11.   Distribution of registered trademark groups among countries of sample firms.
Source: Crunchbase and authors' calculations.

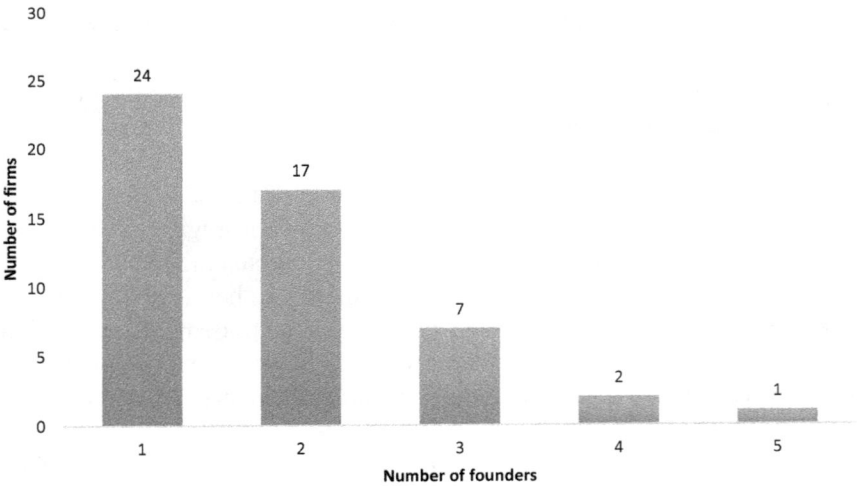

Figure 12.   Firms by number of founders.
Source: Crunchbase and authors' calculations.

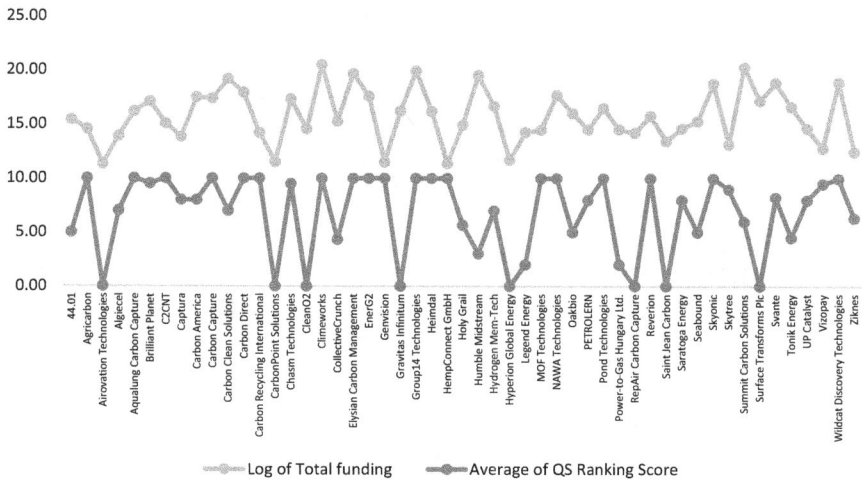

Figure 13.   Average PIS of firms versus log total funding.

Source: Crunchbase and authors' calculations.

Such institutes provide founders with alumni networks, goodwill accumulated by the institute over time, and education. Figure 13 shows line plots for PIS versus the log of total funding, showing comparably simultaneous movements. This reflects the relationship emerging between qualifications from premier institutes and access to funding in CCS.

Last, we explore the founders' highest qualification, which is used for categorization into relevant areas of education, as represented in Figure 14. A plurality of founders has relevant education, with roughly proportional shares of B.S., M.S., and Ph.D. For other educational qualifications, M.S., B.S., and MBAs are present in descending order of prevalence.

4. Results

4.1. *Methodology*

We supplement the foregoing with a two-stage firm-level statistical analysis. The first stage considers the factors affecting access to funding for CCS firms, using ordinary least squares estimation, where the log of total

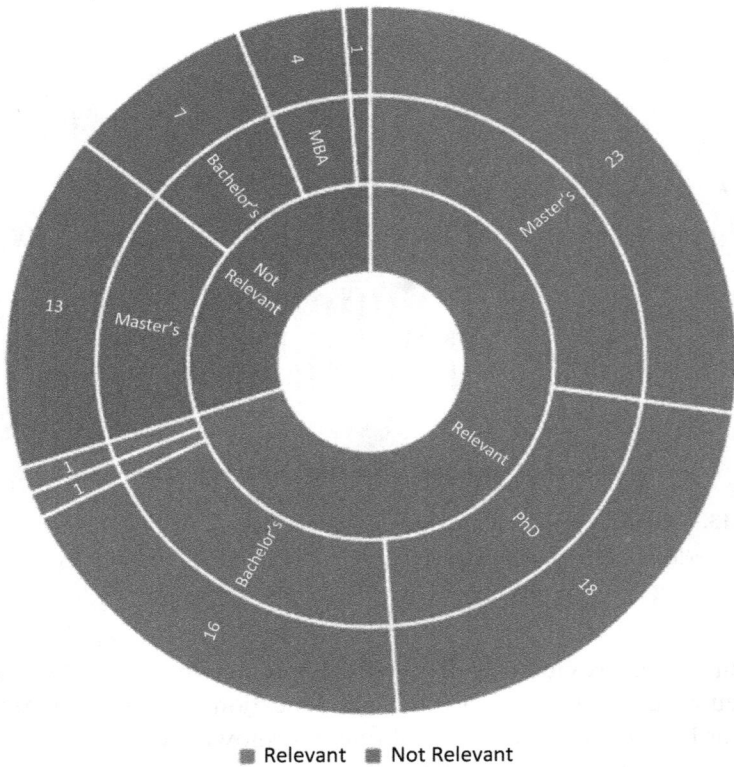

Figure 14.   Distribution of highest educational qualifications of founders of sample firms.
Source: Crunchbase and authors' calculations.

financing is the dependent variable. The full model specification is as follows:

$$
\begin{aligned}
(Funding) = {} & \beta_1\, Patents\, and\, Trademarks_i + \beta_2\, PIS_i \\
& + \beta_3\, Industry\, experience_i \\
& + \beta_4\, Country\, readiness_i + \beta_5\, Specialization_i \\
& + \beta_6\, Prior\, Startup\, Experience_i \\
& + \beta_7\, Founders_i + \beta_8\, Age_i \\
& + \beta_9\, (Patents\, and\, Trademarks_i \times PIS_i) \\
& + \beta_{10}\, (Founders_i \times Industry\, experience_i) + \varepsilon_i
\end{aligned} \tag{1}
$$

$$
\begin{aligned}
Patents\ and\ Trademarks_i = {} & \beta_1\ Funding_i + \beta_2\ PIS_i \\
& + \beta_3\ Industry\ Experience_i \\
& + \beta_4\ Country\ Readiness_i + \beta_5\ Specialization_i \\
& + \beta_6\ Prior\ Startup\ Experience_i \\
& + \beta_7\ Founders_i + \beta_8\ Age_i \\
& + \beta_9\ (Funding_i \times PIS_i) \\
& + \beta_{10}\ (Funding_i \times Specialization_i) \\
& + \beta_{11}\ (Founders_i \times Industry\ experience_i) + \varepsilon_i
\end{aligned}
\tag{2}
$$

where ε_i is the error-term. The second-stage analysis examines factors accounting for higher numbers of patents by startups. We employed negative binomial regression as the total patents and trademarks registered is a count variable. This methodology provides an appropriate measure for determining the causal relationship between dependent and independent variables in the presence of overdispersion in a count-dependent variable. The primary variables of interest are log of total funding to probe the impact of funding on patents and trademark acquisitions, and specialization score, which indicates the role of education in specialized fields relevant to real-world CCS applications. Country readiness score is excluded as it is fundamentally determined by government policy.

4.2. Results

The results of the first-stage analysis are presented in Table 4. The regressions are conducted with interactions among the primary variables of interest with controls and without a constant, as shown in Equation (1). Column (4) gives the coefficients for the full model specification with all interaction terms. Model 1 reports the regression results without any interaction terms, and Models 2 and 3 give results for each interaction separately in the presence of controls. Among major funding drivers, patents and trademarks, premium institute score, industry experience, and country readiness exhibit direct and positive impact, indicating notable roles played by these. However, the interaction terms have insignificant coefficients with contrasting signs in the full model, rendering their interaction effects inconclusive. As expected, more founders significantly correlates with greater funding, which appears reasonable. Although we have not

Table 4.   Regression results.

Variables	Dependent variable: Funding			
	(1) **Funding**	**(2)** **Funding**	**(3)** **Funding**	**(4)** **Funding**
Patent and trademarks	0.054*		0.055*	0.218
	(0.029)		(0.029)	(0.145)
PIS	0.417***	0.421***	0.385***	0.387***
	(0.119)	(0.120)	(0.124)	(0.124)
Industry experience	0.121**	0.124**	0.120**	0.112*
	(0.055)	(0.056)	(0.055)	(0.056)
Country readiness	0.136***	0.137***	0.133***	0.127***
	(0.020)	(0.021)	(0.021)	(0.021)
Specialization	−0.372	−0.328	−0.288	−0.313
	(0.676)	(0.681)	(0.682)	(0.680)
Prior startup experience	0.676	0.656	2.311	2.569
	(0.890)	(0.899)	(1.942)	(1.948)
Founders (in numbers)	1.384*	1.312*	1.666**	1.926**
	(0.740)	(0.747)	(0.798)	(0.827)
Age (in months)	0.0006	0.0007	0.0002	0.0002
	(0.002)	(0.002)	(0.002)	(0.002)
Patents and trademarks *× PIS*		0.002		−0.008
		(0.001)		(0.007)
Founders × industry *experience*			−0.943	−1.059
			(0.996)	(0.997)
Observations	50	50	50	50
R-squared	0.969	0.968	0.969	0.970

Note: Standard errors are in parentheses. *** $p<0.01$, ** $p<0.05$, * $p<0.1$. PIS = premier institute score.

accounted for gender, future studies may examine whether this affects funding as well.

In the second stage, we analyze patents and trademarks registered with negative binomial regression per Equation (2). Table 5 reports the results for the regressions, with columns 1, 2, and 3 reporting coefficients

Table 5. Negative binomial regression results.

Variables	(1) Patents and trademarks	(2) Patents and trademarks	(3) Patents and trademarks	(4) Patents and trademarks
Funding			0.188	0.085
			(0.171)	(0.162)
PIS		0.162	0.009	−1.038
		(0.120)	(0.143)	(0.705)
Industry experience	0.033	0.045		0.007
	(0.045)	(0.044)		(0.048)
Country readiness	−0.017	−0.018	−0.054	−0.019
	(0.018)	(0.019)	(0.036)	(0.037)
Specialization	1.588***		1.589*	2.449
	(0.593)		(0.883)	(3.157)
Prior experience	−1.773**	−1.807**	−0.621	−0.541
	(0.894)	(0.897)	(2.616)	(2.215)
Founders (in numbers)	−0.745	−0.514		−0.765
	(0.709)	(0.733)		(0.689)
Age (in months)	0.002	0.002	0.002	−0.0004
	(0.003)	(0.003)	(0.004)	(0.003)
Funding × PIS	0.014**			0.069
	(0.006)			(0.043)
Funding × specialization		0.100***		−0.068
		(0.032)		(0.183)
Founders × industry experience			−0.417	−0.228
			(1.534)	(1.117)
Observations	50	50	50	50

Note: Standard errors are in parentheses. *** $p<0.01$, ** $p<0.05$, * $p<0.1$. PIS = premier institute score.

for the model specified by separating the interaction terms, and column 4 reports the results for the full-model specification with interactions between the primary variable of interest. Funding has insignificant positive coefficients, while specialization has significant coefficients, except in the case of the full model specification, where none of the variables or interaction terms have a significant coefficient. The interaction between funding and specialization has a significant positive coefficient, highlighting that specialized educational qualification is a crucial factor affecting the patents and trademarks generation for the firm. Funding is the driving force, however, as its lack would hinder the optimal application of the specialized skillset. While similar inferences can be drawn from the significant positive coefficients for the interaction between funding and PIS, PIS alone is insignificant to patent and trademark acquisitions by firms.

5. Discussion and Conclusion

This study offers the following lessons for Southeast Asia. First, the first-stage analysis suggests a concentration of CCUS startups in the US, Canada, and the UK. This means that ASEAN member states must explore technological collaboration focusing on endogenous capabilities and human resources. This is recommended because these countries have designed policies and encouraged private firms and investment in CCUS development and deployment. The analysis also reveals the status of CCUS and how its deployment can be economical, especially for hazardous industries and manufacturing units where carbon capture may make a big difference. Southeast Asian economies are also middle-class, and some carbon products may be used for household applications and product development. Livelihood and sustainability are other dimensions that CCUS offers. Its deployment in the cement, steel, and petroleum industries may make a difference.

The second-stage analysis focuses on the role of the private sector and how some public sector policies designed to expand CCUS innovations may create a knowledge economy and better employment opportunities. Large-scale data analysis by machine learning, new methods of capturing carbon by startups, the demographic advantage of the same, new carbon-based products, and many patent registrations are some of the salient lessons for Southeast Asian economies. There are enormous learning opportunities for ASEAN as these economies are at the initial stages of CCUS promotion and deployment. Startups' diversification and efficiency efforts may certainly add value to the CCUS development and deployment efforts in Southeast Asia and ASEAN. Although the empirical design of this chapter does not show significant CCS drivers, it suggests PIS, patents and trademarks, country readiness, and industry experience play a significant role. From ASEAN's perspective, these inferences speak volumes about promoting a startup ecosystem, creating a knowledge-driven economy, and promoting cross-country knowledge collaboration. And as noted previously, the composition of founders matters for funding. In this regard, specialization appears to be the key driver of patents and trademarks, showing the role of research and development environments.

Concerning best practices, the authors believe that private sector promotion through venture capital, investment directed through multilateral organizations, and efficiency and carbon-based product development initiatives are some of the relevant avenues of focus for the ASEAN economies. These implications are crucial as the study focuses on endogenous CCUS development and deployment capabilities. The US, UK, and Canada offer tremendous lessons. ASEAN member states must focus on education and cross-country collaborations. In this regard, a dedicated CCUS fund for university-level incubators and accelerators may give new life to the low-carbon transition. Cleantech is already picking up in ASEAN, and promoting CCUS-focused startups may create a skilled workforce for the next industrial revolution.

The major limitation of this study has been the availability of data. Future studies may create a larger dataset to validate our findings. Panel data models may robustly capture fixed effects and heterogeneity.

References

Anderson, S. and Newell, R. (2004). Prospects for carbon capture and storage technologies. *Annual Review of Environment and Resources*, 29, 109–142.

Bui, M., Adjiman, C. S., Bardow, A., Anthony. E. J., Boston, A., Brown, S., ... and Mac Dowell, N. (2018). Carbon capture and storage (CCS): The way forward. *Energy & Environmental Science*, 11(5), 1062–1176.

Chang, Y. and Phoumin, H. (2022). Curtailed electricity and hydrogen in Association of Southeast Asian Nations and East Asia Summit: Perspectives form an economic and environmental analysis. *International Journal of Hydrogen Energy*, 47(58), 24548–24557.

IEA. (2021). *World Energy Outlook 2021*, IEA, Paris. https://www.iea.org/reports/world-energyoutlook-2021.

Fennell, P. S., Florin, N., Napp, T., and Hills, T. (2012). CCS from Industrial Sources. *Sustainable Technologies, Systems and Policies, Carbon Capture and Storage*, Workshop 17.

Guo, J. X. and Huang, C. (2020). Feasible roadmap for CCS retrofit of coal-based power plants to reduce Chinese carbon emissions by 2050. *Applied Energy, 259*, 114112.

Hills, T. P., Sceats, M. G., and Fennell, P. S. (2019). Applications of CCS in the Cement Industry. In *Carbon Capture and Storage* (pp. 315–352). London: Royal Society of Chemistry.

Kearns, D., Liu, H., and Consoli, C. (2021). Technology readiness and costs of CCS. Global CCS Institute, Australia.

Kimura, S., Shinchi, K., Kawagishi, S., and Coulmas, U. (2021a). *Study on the Potential for the Promotion of Carbon Dioxide Capture, Utilisation, and Storage in ASEAN Countries: Current Situation and Future Perspectives.* ERIA Research Project Report FY2020 No. 21, Economic Research Institute for ASEAN and East Asia (ERIA).

Kimura, S., Shinchi, K., Coulmas, U., and Saimura, A. (2021b). *Study on the Potential for Promoting Carbon Dioxide Capture, Utilisation, and Storage (CCUS) in ASEAN Countries Vol. II.*, ERIA Research Project FY2021 No. 25. Economic Research Institute for ASEAN and East Asia (ERIA).

Leeson, D., Mac Dowell, N., Shah, N., Petit, C., and Fennell, P. S. (2017). A Techno-economic analysis and systematic review of carbon capture and storage (CCS) applied to the iron and steel, cement, oil refining and pulp and paper industries, as well as other high purity sources. *International Journal of Greenhouse Gas Control, 61*, 71–84.

Li, J., Tharakan, P., Macdonald, D., and Liang, X. (2013). Technological, economic and financial prospects of carbon dioxide capture in the cement industry. *Energy Policy, 61*, 1377–1387.

Li, W., Jia, Z., and Zhang, H. (2017). The impact of electric vehicles and CCS in the context of emission trading scheme in China: A CGE-based analysis. *Energy, 119*, 800–816.

Liang, X. and Li, J. (2012). Assessing the value of retrofitting cement plants for carbon capture: A case study of a cement plant in Guangdong, China. *Energy Conversion and Management, 64*, 454–465.

Lilliestam, J., Bielicki, J. M., and Patt, A. G. (2012). Comparing carbon capture and storage (CCS) with concentrating solar power (CSP): Potentials, costs, risks, and barriers. *Energy Policy, 47*, 447–455.

Liu, H. and Gallagher, K. S. (2010). Catalyzing strategic transformation to a low-carbon economy: A CCS roadmap for China. *Energy Policy, 38*(1), 59–74.

Nepal, R., Phoumin, H., and Khatri, A. (2021). Green technological development and deployment in the association of Southeast Asian economies (ASEAN) — At crossroads or roundabout? *Sustainability, 13*(2), 758.

Paltsev, S., Morris, J., Kheshgi, H., and Herzog, H. (2021). Hard-to-Abate Sectors: The role of industrial carbon capture and storage (CCS) in emission mitigation. *Applied Energy, 300*, 117322.

Phoumin, H., Kimura, F., and Arima, J. (2020). Potential renewable hydrogen from curtailed electricity to decarbonize ASEAN's emissions: Policy implications. *Sustainability, 12*(24), 10560.

Poudyal, L. and Adhikari, K. (2021). Environmental sustainability in cement industry: An integrated approach for green and economical cement production. *Resources, Environment and Sustainability, 4*, 100024.

Rootzén, J. and Johnsson, F. (2015). CO_2 emissions abatement in the Nordic carbon-intensive industry–an end-game in sight? *Energy, 80*, 715–730.

Stokke, R., Kristoffersen, F. S., Stamland, M., Holmen, E., Hamdan, H., and De Boer, L. (2022). The role of green public procurement in enabling low-carbon cement with CCS: An innovation ecosystem perspective. *Journal of Cleaner Production*, 132451.

Turan, G., Zapantis, A., Kearns, D., Tamme, E., Staib, C., Zhang, T., ... and Steyn, M. (2021). *Global status of CCS 2021. CCS Accelerating to Net Zero.* Melbourne: Global CCS Institute.

Part 2

Chapter 5

Nature-Based Carbon Capture through Forest Management in the Mekong Region of Southeast Asia

Nophea Sasaki[*,‡] and Sharaniya Vijitharan[†,§]

Natural Resources Management, School of Environment, Resource, and Development, Asian Institute of Technology, Pathum Thani, Thailand

†*Department of Bio-Science, Faculty of Applied Science, University of Vavuniya, Vavuniya, Sri Lanka*

‡*nopheas@ait.ac.th*

§*sharaniya@vau.ac.lk*

Abstract

This research examines the carbon capture potential and timber production outcomes of conventional logging (CVL) and reduced impact logging (RIL) in the Mekong region of Southeast Asia from 1990 to 2050, showing that CVL resulted in approximately 4590.9 $TgCO_2$ in carbon emissions during the study period, while RIL demonstrated a notable annual reduction of 10.9 $TgCO_2$ in carbon emissions, alongside sustainable annual production of 11.4 million m^3 of wood. Plantation forests (PF) contributed to an annual carbon removal of 58.69 $TgCO_2$ over the course of the study. Implementing forest management practices in the Mekong may also generate significant carbon revenues, estimated at USD 3,500.07 million within the Paris Agreement period 2020–2030, contingent upon prevailing carbon pricing mechanisms. Embracing sustainable forest management (SFM) and restoration strategies, with emphasis on RIL, may drive improvements in quality of wood products, waste reduction, and natural carbon capture enhancement.

Keywords: Nature-based carbon capture, forest, CVL, RIL, Mekong region.

1. Introduction

Nature-based carbon capture in forests is widely recognized as a crucial measure for mitigating climate change and promoting sustainable development (World Bank, 2022). Tropical forests face severe challenges from deforestation and degradation, leading to substantial carbon emissions, biodiversity loss, poverty, and other consequences (Bösch, 2021). Nature-based solutions have emerged as essential tools for climate change mitigation and adaptation (Sasaki, 2021; Sasaki *et al.*, 2021). Achieving this involves forest conservation, restoration, and sustainable management (Player and Taber, 2021). Implementing carbon capture and storage through nature-based solutions is also imperative for decreasing carbon emissions and enhancing carbon recycling pathways. In line with the objectives of the Paris Agreement, numerous national and international initiatives have prioritized Reducing Emissions from Deforestation and forest Degradation (REDD+), as well as promoting forest conservation, sustainable forest management (SFM), and enhancement of forest carbon stocks (UNFCCC, 2010).

During the 26th Conference of the Parties (COP 26), global leaders agreed to restore all degraded forests by 2030, aiming to increase nature-based carbon capture while improving forest ecosystem services to support economic growth (Taylor, 2021). According to the International Tropical Timber Organization (ITTO, 2005), SFM is defined as "the process of managing permanent forestland to meet one or more defined management goals regarding the production of a steady stream of desired forest products and services without causing a significant decrease in its inherent values and future productivity or having a significant negative impact on the physical and social environment." This definition highlights the importance of sustainable forest management practices in balancing economic productivity with the preservation of environmental values and long-term forest productivity.

Implementing SFM, particularly as part of REDD+, plays a vital role in preventing carbon loss in forests, increasing carbon storage in wood, and simultaneously ensuring wood supplies and generating revenues for governments in developing nations (Sasaki *et al.*, 2016). SFM, as highlighted by Piponiot *et al.* (2018), enables maintaining timber supplies until new trees reach harvestable ages (Sasaki *et al.*, 2021), reduces forest fires, preserves carbon stocks, and decreases emissions (Sasaki *et al.*, 2016). Adopting an integrated management strategy accompanied by appropriate incentives is essential for effective management of crucial forest resources. Recent studies have emphasized environmentally friendly management practices for tropical forests, focusing on adopting reduced-impact logging (RIL) instead of conventional logging (CVL). RIL practices prevent physical and environmental damage in addition to other long-term benefits (Boltz *et al.*, 2003; Holmes *et al.*, 2002). Plantation forests also play a significant role in mitigating climate change, especially in tropical regions with substantial carbon sequestration potential (Smith *et al.*, 2014; FAO, 2016). Therefore, exploring integrated approaches to sustainable forest management that facilitate nature-based carbon capture and wood production through improved logging practices is crucial, particularly in developing regions.

This study focuses on forests in the Mekong region of Southeast Asia due to the region's heavy reliance on their timber to support economic growth (Yasmi *et al.*, 2017). Encompassing Thailand, Laos, Cambodia, Myanmar, and Vietnam, the Mekong is renowned as one of the world's most significant biodiversity hotspots. Its forests hold great value for the region's development, playing a crucial role in wood production (WWF, 2018; Yasmi *et al.*, 2017). Between 1990 and 2015, Mekong forest cover declined 5.1%, chiefly in Cambodia, Myanmar, and Thailand, while Laos and Vietnam showed increases, primarily driven by expanded commercial plantations rather than natural forests (RECOFTC, 2020). In developing regions, illegal and unsustainable logging practices have severe impacts on forest sustainability (Vasco *et al.*, 2017; Bösch *et al.*, 2018). Similarly, forest degradation in the Mekong is exacerbated by agricultural expansion, plantations, legal and illegal logging, and other developmental activities (WWF, 2018). The prevalence of poor-quality logging practices has significantly diminished the present and future value of the Mekong's forests economically and ecologically alike. It is projected that approximately 15–30 million ha of forests will be lost by 2030 without intervention, emphasizing the urgent need for SFM and effective forest governance

(WWF, 2018), as SFM has not been widely implemented for production purposes (FAO, 2011).

The objective of this study is to assess changes in carbon stocks, timber production, and carbon capture through forest management under the logging practices of CVL and RIL, as well as carbon capture through plantation forests in the Mekong from 1990 to 2050, with emphasis on emission reduction during the implementation period of the Paris Agreement, 2020–2030.

2. Material and Methods

2.1. *Mekong forest resources*

The Mekong comprises Thailand, the Lao People's Democratic Republic (PDR), Cambodia, Myanmar, and Vietnam. According to the Forest Resources Assessment by the FAO (FAO, 2020), the total Mekong forest area in 2020 was 87,723,850 ha, of which 87.8 percent consisted of naturally regenerating forests, with the remaining 12.2 percent designated plantation forest. The naturally regenerating forests were classified as production forest, protection forest, conservation forest, forest for social services, multiple-use forest, and other forests (FAO, 2020). For the purposes of this study, we grouped all remaining forest areas as protection forest (PRF) except for production forest (PDF). Our analysis focused on the PDF and plantation forest (PF), as shown in Figure 1.

Selective logging techniques were employed in managing the PDF, in which timber is extracted over a 30-year cutting cycle (Piponiot *et al.*, 2019a, 2019b). CVL is the most common form of selective logging in Southeast Asia, prioritizing immediate financial gains from commercial logging over careful planning, mapping, and training. It was not anticipated that the PRF would be logged, as it is primarily managed for such

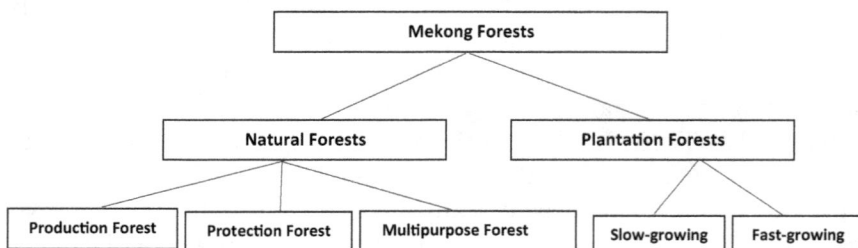

Figure 1. Forest land use types and management systems in the Mekong.

purposes as biodiversity protection, watershed protection, and creative endeavors, rather than commercial logging. Plantation forests can be classified into two types based on their cutting rotations: fast-growing plantation forest (FPF) and slow-growing plantation forest (SPF). Approximately 47 percent of the total plantation forest area in Southeast Asia is covered by fast-growing plants, with the remainder consisting of slow-growing species (Sasaki *et al.*, 2009; FAO, 2000). In FPF, clear-cutting is commonly employed, allowing for tree harvests approximately 5–15 years after planting (FAO, 2006). In the case of SPF, the cutting rotations range from 25 to 45 years, depending on the species and location of the trees (Nizami *et al.*, 2014). Fast-growing species, such as eucalyptus and acacia, along with other non-native species, are commonly found in FPF (FAO, 2001; Huy, 2004; Calvo-Alvarado *et al.*, 2007; Plath *et al.*, 2011). Slow-growing tree species, including *Tectona grandis*, *Dipterocarps*, shorea, and others, are frequently found in SPF (Pandey and Brown, 2000). For our analysis, we assumed that clear-cutting and selective logging were employed in managing the PF, with SPF and FPF being harvested over 10-year and 30-year cutting cycles, respectively.

2.2. *Forest cover model for the Mekong*

Land use models were developed to forecast future availability of land for production and plantation forests in the Mekong, as depicted in Figure 2. This study adopted an approach similar to the classification by Sasaki *et al.* (2009). In our model, as shown in Figure 1, the PDF encompassed the areas designated for production, multiple-use, and other forest purposes. These lands were set aside for activities such as farming, urbanization, commercial logging, and land development. On the other hand, protected and conservation forests were categorized as areas designated for protection. The protected forest remained constant throughout the modeling process, as mandated by law and definition. It is important to note that clearing the PDF is strictly prohibited. In this study, we assumed that plantation forests would partially replace deforested lands. Table 1 presents total forest area categorized by forest type in the Mekong.

Between 1990 and 2020, the natural Mekong forest declined continuously at an annual rate of 519,925.3 ha. By contrast, the area covered by plantation forest grew over the same period, from 4,168,990 ha in 1990 to 10,688,680 ha in 2020, per Table 1. A retrospective approach was employed to project forest cover changes utilizing available data (IPCC, 2006). Table 1

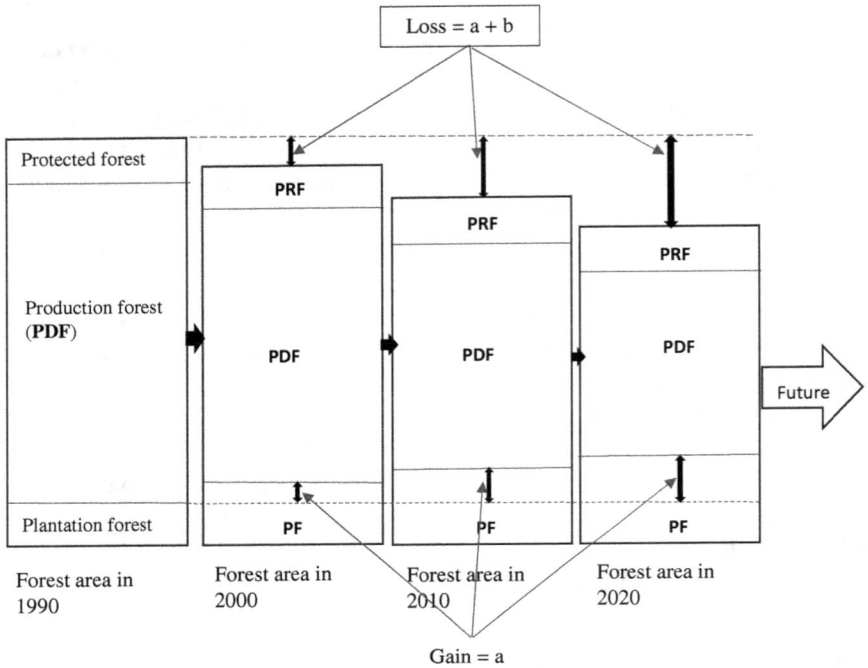

Figure 2.   Illustration of forest land use change model in the Mekong.
Source: Modified from Sasaki *et al.* (2009).

Table 1.   Areas of natural and plantation forests in the Mekong, 1990–2020.

	Forest cover in the Mekong region (million ha)			
	Natural forest		**Plantation**	**Total forest**
Year	**Production forest**	**Protected forest**	**forest**	**(million ha)**
1990	71.3	21.3	4.4	97.1
2000	61.7	26.6	5.9	94.1
2010	53.0	31.0	8.4	92.4
2020	45.8	31.2	10.7	87.7

indicates that certain deforested areas may have been partially converted to plantation forests, fast-growing and slow-growing alike, as well as other land uses (Stibig *et al.*, 2007). Therefore, for our analysis, we considered only deforested area that was subsequently converted to plantation forests.

Table 2. Initial values and rates of change for Equations (1) and (3).

PDF			PF		
Initial value (M ha)	$a + b$	Rate of change	Initial value (M ha)	a	Rate of change
71.34	−0.0148	−1.48	4.44	0.0023	0.23

To calculate the changes in the conversion from production forest to plantation forest, we employ Equations (1)–(3) as proposed by Kim Phat *et al.* (2004):

$$\frac{dPDF(t)}{dt} = (a + b) \times PDF(t) \tag{1}$$

$$\frac{dPRF(t)}{dt} = 0 \tag{2}$$

$$\frac{dPF(t)}{dt} = a \times PDF(t) \tag{3}$$

where $PDF(t)$ represents the area of production forests at time t (in hectares), $PF(t)$ represents the area of plantation forest at time t (in hectares), $(a + b)$ represents the rate of change for the production forest (%), and (a) represents the rate of change of PDF to PF.

Based on the data in Table 1, linear regression analysis was conducted to determine the values of $(a + b)$ and (a). The initial values and parameter values E or Equations (1) and (2) are in Table 2.

2.3. Models for carbon stocks and timber productions from production forest under CVL and RIL

The literature recognizes the aforementioned RIL and CVL primary selective logging categories. While CVL is widely practiced in tropical regions, it has drawbacks, such as the aforementioned absence of logging planning, mapping, and trained workers, resulting in damage to remaining forest stands, substantial wood waste, and increased carbon emissions (Brown *et al.*, 2011; Ellis *et al.*, 2019; Pearson *et al.*, 2014). On the other hand, RIL represents a sustainable harvesting method that aims to reduce carbon emissions and minimize damage to the remaining forests caused by logging activities (Sasaki *et al.*, 2012).

The objective of this study was to assess changes in carbon stocks and emission reductions in the production forest under CVL and RIL. A modeling approach was employed to analyze the period 1990–2050, which encompasses two cutting cycles. To estimate and compare carbon stocks, timber production, and emission reductions, the following equations, modified from Kim Phat *et al.* (2004), were utilized under the respective logging systems. A detailed description of the symbols, justification, and references for predicting carbon stocks in the PDF can be found in Table 3.

$$\frac{dCS_{PDF_i}(t)}{dt} = MAI - LM_i(t) - H(t) \times BEF \qquad (4)$$

$$H(t) = \frac{f_M \times f_H}{1-r} \times \frac{CS_{i(t)}}{T_c \times BEF} \qquad (5)$$

Initial carbon stocks per unit area for the natural forest were determined by calculating the weighted average of forest carbon stocks across the Mekong (FAO, 2020), yielding a result of 63.09 MgCha^{-1} or tCha^{-1}. This study focused solely on the aboveground carbon pool, disregarding other carbon pools due to substantial variations among different forest types (Dar and Sundarapandian, 2015).

We utilized Equations (6)–(14) under both logging systems to estimate quantities of various wood components, including wood products (WP), waste (WAS), logging mortality (LM), end-use wood products (EWP), and end-use wood wastes (EWAS):

$$WP_i(t) = (1 - s_i) \times H_i(t) \qquad (6)$$

$$WAS_i(t) = H_i(t) - WP_i(t) \qquad (7)$$

$$LM_i(t) = \alpha \times H_i(t) \qquad (8)$$

$$EWP_i(t) = a_i(t) - WP_i \qquad (9)$$

$$EWAS_i(t) - EWP_i(t) - \text{EWP}_i(t) \qquad (10)$$

Achieving sustainable forest management and climate change mitigation becomes challenging if the preservation of a long-term sustainable supply of wood is not one of the management objectives. In this analysis, we compared end-use wood products (EWP) generated through RIL with

Table 3.   Symbols and references used in predicting carbon stocks and timber production in PDF.

Symbol	Description	Value CVL	Value RIL	Reference
CS_{PDF_i}	Initial carbon stocks (MgC/ha)	63.09		
CS_{PDF_i}	Aboveground carbon stocks under CVL and RIL (Average carbon stocks calculated from the countries in the Mekong region) (MgC/ha)			FAO (2020)
MAI	Mean Annual Increment (MgC/ha yr)	0.66		Butarbutar *et al.* (2019)
$LM_i(t)$	Loss of carbon caused by logging-induced mortality (MgC/ha)			
$H(t)$	Carbon in the harvested wood (MgC/ha), which can be converted to timber volume (m³) by $H(t)/(WD \times CF \times BEF)$			
WD	Wood density (Mg m⁻³)			Brown (1997)
CF	Carbon fraction in the dry wood	0.47		IPCC (2006)
BEF	Biomass expansion factor	1.74		Brown (1997)
f_M	Proportion of mature trees (above the optimum limit of diameter for harvesting)	0.43		Kim Phat *et al.* (2004)
f_H	Legal rate of harvesting (allowed legally to harvest by the government)	0.30		Kim Phat *et al.* (2004)
R	Illegal logging (average global rate was 0.5)	0.50		Lowe *et al.* (2016)
T_C	Cutting cycle in years	30		Khai *et al.* (2020)
a	Proportion of useless wood remaining after subtracting losses from logging, skidding, and transportation-related damage	0.3	0.1	Sasaki *et al.* (2012)
α	Proportion of trees killed by log skidding and logging (LD in the CVL can be reduced from 40 percent to 14 percent by executing proper planning in logging practice)	0.40	0.14	Matangaran *et al.* (2019); Butarbutar *et al.* (2019)
s	Wood processing efficiency	0.50	0.40	Sasaki *et al.* (2012)

Source: Adopted from Sasaki *et al.* (2012), unless otherwise stated.

those produced under CVL, which served as the wood production baseline. Our goal was to ensure that the EWP values obtained from both logging methods were equivalent. We used the following equations to calculate EWP under CVL and RIL:

$$EWP_{CVL}(t) = (1 - \alpha_{CVL}) \times WP_{CVL}(t) \qquad (11)$$

$$EWP_{RIL}(t) = (1 - \alpha_{RIL}) \times WP_{RIL}(t) \qquad (12)$$

$$EWP_{RIL}(t) = EWP_{CVL}(t) \qquad (13)$$

or

$$H_{RIL}(t) = \frac{(1 - a_{CVL})(1 - s_{CVL})}{(1 - a_{RIL})(1 - s_{RIL})} \times H_{CVL}(t) \qquad (14)$$

WP, WAS, LM, EWP, and EWAS are measured in MgCha^{-1} per year. However, timber production and wood wastes are typically measured in cubic meters (m^3). To calculate the total amount of wood products for the scenario mentioned, the corresponding variables are multiplied by the area of the PDF.

2.4. *Baseline emission level, project emission level, and carbon credits*

Baseline emission level (BEL) represents emissions under CVL, which serves as a reference for emissions in the absence of project activities. Conversely, project emission level (PEL) corresponds to emissions under RIL, reflecting emissions resulting from the implementation of project activities. By comparing BEL with PEL, we can estimate the emission reduction achieved by transitioning from CVL to RIL. This reduction in carbon emissions is eligible for performance-based incentives under the REDD+ scheme of the United Nations Framework Convention on Climate Change (UNFCCC).

$$BEL(t) = [TCS_{CVL}(t2) - TCS_{CVL}(t1)] \times 44/12 \qquad (15)$$

$$PEL(t) = [TCS_{RIL}(t2) - TCS_{RIL}(t1)] \times 44/12 \qquad (16)$$

$$TCS(t) = CS_i(t) \times PDF \qquad (17)$$

$$CC(t) = [BEL(t) - PEL(t)] \times [1 - L(t)] - EP(t) \qquad (18)$$

where BEL represents emissions under CVL, in teragrams of CO_2 per year (TgCO$_2$ year^{-1}), with 1 TgCO$_2$ = 10^6 MgCO$_2$; PEL corresponds to emissions under RIL, again, in TgCO$_2$ year^{-1}; Leakage (L) refers to emissions outside the project boundaries, also in TgCO$_2$ year^{-1}; and EP(t) represents emissions from project activities, such as law enforcement and use of motorbikes by forest rangers. The molecular weight of CO_2 to carbon is 44/12. Carbon credits (CC) are earned based on achieved emissions reductions.

Leakage (L) varies significantly between different sites and is thus challenging to measure. In this analysis, we assumed a leakage rate of 0.30 (Sasaki *et al.*, 2012). It should be noted that EP(t) was assumed to be equivalent for logging and wood transportation under both CVL and RIL. EP(t) is estimated at less than 10 percent (Ty *et al.*, 2011), which was excluded pursuant to UNFCCC rules (UNFCCC, 2011) which state that emissions that account for 10 percent or less of the total may be thus excluded.

2.5. *Models for carbon stocks and carbon capture in plantation forests*

Clear-cutting was applied to FPFs when they reached their cutting rotation of 10 years, while SPFs were clear-cut at their cutting rotation of 30 years. The following equations were utilized to calculate various parameters associated with PF management, including total harvested carbon, total carbon stocks, carbon sequestration or removals, and carbon revenue:

$$H_{PF}(t) = \frac{\text{CS}_{PF}(t)}{\text{T}_R} \tag{19}$$

Total harvested carbon can be estimated by

$$TH_{PF}(t) = H_{PF}(t) \times PF(t) \tag{20}$$

Total carbon stocks in FPF and SPF can be computed by

$$TCS_{FPF}(t) = \left[\frac{\text{CS}_{PF}(t) \times \text{PF}(t) - TH_{PF}(t)}{1,000,000} \right] \times 0.47 \tag{21}$$

$$TCS_{SPF}(t) = \left[\frac{\text{CS}_{PF}(t) \times \text{PF}(t) - TH_{PF}(t)}{1,000,000} \right] \times 0.53 \tag{22}$$

Carbon sequestration in FPF and SPF can be calculated by

$$CR_{FPF}(t) = [TCS_{FPF}(t2) \times TCS_{FPF}(t1)] \times \frac{44}{12} \qquad (23)$$

$$CR_{SPF}(t) = [TCS_{SPF}(t2) \times TCS_{SPF}(t1)] \times \frac{44}{12} \qquad (24)$$

where $H_{PF}(t)$ is harvested carbon in FPF and SPF at time t (MgCha^{-1}); T_R: Cutting rotation (yr); $CS_{PF}(t)$ is carbon stock in plantation forest (44.5 MgCha^{-1}, FAO, 2020), which remains constant (IPCC, 2006); $TH_{PF}(t)$ is total harvested carbon (MgC); $TCS_{FPF}(t)$ is total carbon stocks remaining in FPF at time t (TgC); $TCS_{FPF}(t)$ is total carbon stocks remaining in SPF at time t (TgC); 1,000,000 is the conversion factor from MgC to TgC; 0.47 (47 percent) and 0.53 (53 percent) is share of FPF and SPF in total plantation forest; $CR_{SPF}(t)$ is carbon removal in SPF; and $CR_{FPF}(t)$ carbon removal in SPF at time t.

Total carbon credits can be estimated by summing emissions reductions from PDF and carbon removals from PF, collectively referred to as carbon capture through forest management. Carbon revenues were computed by multiplying carbon credits and carbon price, 1 ton CO_2 = USD 10, as per the price reported in the World Bank Report (2020).

3. Results and Discussion

3.1. *Changes in forest area in the Mekong*

We conducted an assessment and projection of the changes in production and plantation forest cover in the Mekong, 1990–2050, spanning 60 years, equivalent to two cutting cycles of 30 years each. Our analysis focused on production forest and plantation forest to evaluate forest cover changes, carbon stocks, emissions reduction potential, and biomass. By employing a forest land use model, we aimed to develop conservation and sustainable management plans for the region's forest resources.

Our analysis revealed a significant decrease in the area of PDF from 71.34 million ha in 1990, with a lower limit of 62.01 million ha and an upper limit of 80.67 million ha, to 29.43 million ha in 2050, with a lower limit of 20.10 million ha and an upper limit of 38.76 million ha, per Figure 3. This decline was characterized by an annual change of −0.70 million ha and a rate of change of −0.98 percent, per Table 4.

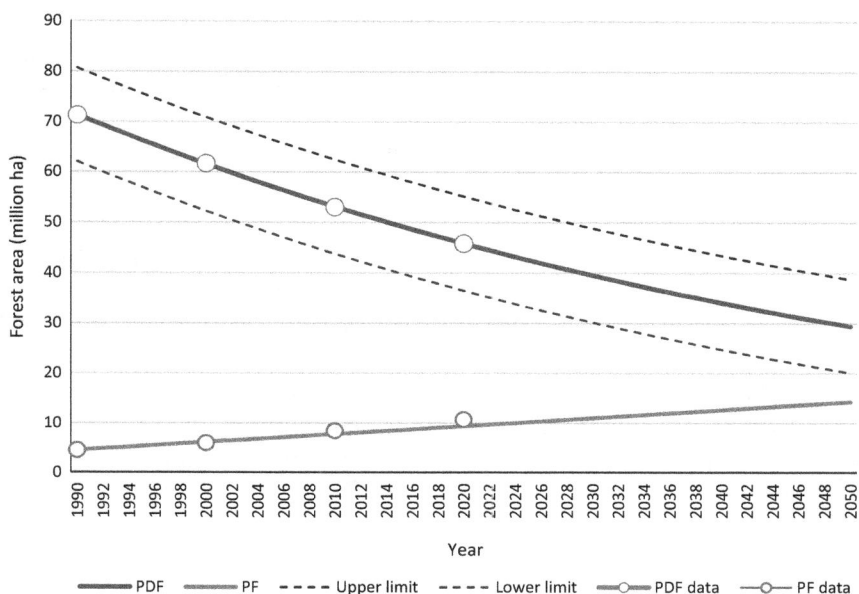

Figure 3. Projected areas of production forest and plantation forest, 1990–2050.

Table 4. Annual rates of Mekong forest cover change, 1990–2050.

Time frame	Production forest (Mha)	Production forest (%)	Plantation forest (Mha)	Plantation forest (%)	Total (Mha)	Total (%)
1990–2020	−0.85	−1.19	0.21	4.69	−0.64	−0.85
1990–2050	−0.70	−0.98	0.16	3.69	−0.53	−0.71

Notably, the area of PF is projected to increase from 4.44 million ha in 1990 to 14.29 million ha in 2050. The average annual growth rate of PF between 1990 and 2050 is estimated to be 0.16 million ha, or 3.69 percent. In 1990, PF covered approximately 5 percent of the total forest area, which is forecast to reach 18 percent by 2050. Conversely, PDF, which accounted for 74 percent of the total forest area in 1990, is projected to decrease to 37 percent by 2050.

Overall, our findings indicate that the Mekong is likely to lose approximately 0.53 million ha per year of natural forest between 1990 and 2050. While expanded PF partially offsets this loss, it is concerning that the pressure on natural forests remains high. Agriculture and illegal and unsustainable logging continue to be the primary drivers of deforestation in the Mekong (Costenbader *et al.*, 2015; Yesmi *et al.*, 2017). These findings highlight the urgent need to address these drivers and implement effective conservation and sustainable management strategies to protect the region's valuable natural forests.

According to the 2015 Global Forest Resource Assessment by the Food and Agriculture Organization, the Mekong was reported to have a total forest area of 88.4 million ha. However, in 2020, the region lost approximately 1.1 million ha, with illegal logging emerging as a major destructive force. The COVID-19 pandemic further exacerbated the situation, as corruption and weak law enforcement contributed to accelerating illegal logging activities (Mongabay, 2021). It is crucial for relevant organizations to prioritize and actively promote SFM practices to protect these forests and mitigate carbon emissions.

Encouraging the transition to responsible logging practices is thus essential in safeguarding the region's forests and mitigating the adverse environmental impacts (MacDicken *et al.*, 2015; Sloan and Sayer, 2015). By prioritizing SFM, it is possible to protect the valuable forest resources of the Mekong and ensure their long-term viability.

3.2. *Harvested timber, wood products, and wastes*

Achieving SFM is essential to ensuring a long-term and successful wood supply in the market. This study confirms that the choice of logging practices, specifically switching from CVL to RIL, can significantly impact harvested timber volumes. Our models demonstrate that in 1990, CVL required harvesting of approximately 33.14 million m^3 of wood, while RIL only needed to harvest 17.18 million m^3 to meet the market's demand of 6.87 million m^3, per Figure 4. Between 1990 and 2050, on average, CVL and RIL are projected to harvest 22.19 million m^3 and 11.41 million m^3 of wood, respectively, to produce 4.56 million m^3 of end-use wood products, again as Figure 4.

During the same period, CVL generated 9.51 million m^3 of wood waste, including logging and wood wastes. By contrast, RIL produced only 1.27 million m^3 of wood waste, resulting in an average reduction of approximately 8.24 million m^3 per year, per Figure 5. Wood waste under

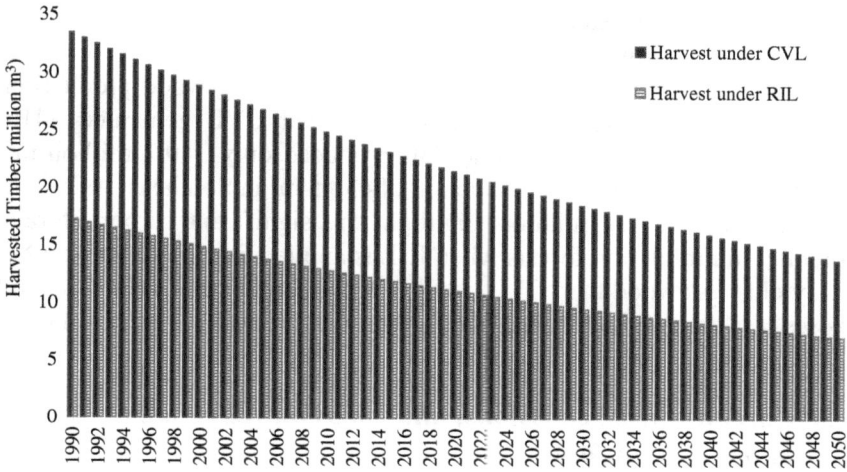

Figure 4. Total harvested timber volume under the CVL and RIL in the Mekong, 1990–2050.

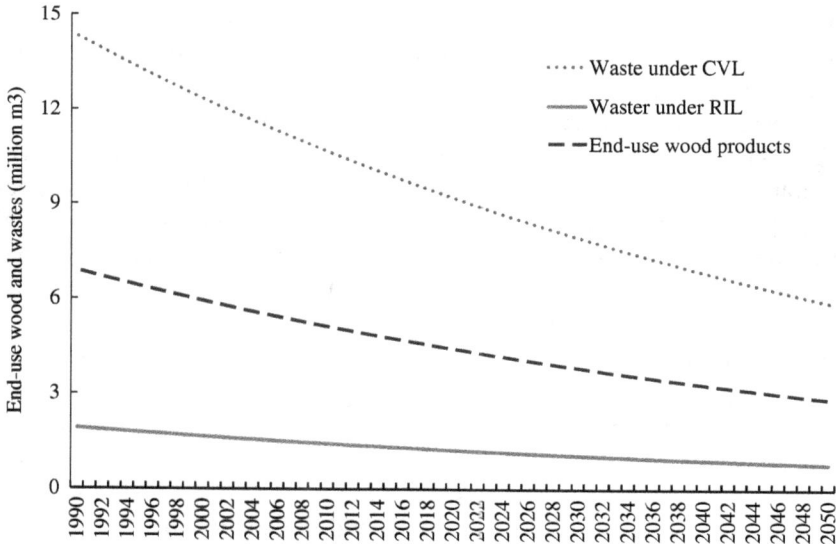

Figure 5. End-use wood products and wastes under the CVL and RIL, 1990–2050.

CVL primarily consisted of sawmill waste and a substantial portion of logging waste, including careless logging, skidding, trimming, and hauling. Although limited studies exist regarding Mekong wood production, research by Sasaki *et al.* (2012) reported that CVL generated 741.6 million m^3 of wood waste, while RIL produced only 340.2 million m^3, leading to an average annual reduction of 431.5 m^3 in Southeast Asia from 2010 to 2060. These findings highlight the significant carbon storage potential in production forests when RIL is applied for timber harvesting. RIL yields less wood than CVL, resulting in greater carbon retention, and causes less damage to residual stands than CVL (Sasaki *et al.*, 2021).

3.3. *Carbon stocks, emissions, and emission reductions in production forest*

The carbon stocks in the production forest of the Mekong region under two logging practices, CVL and RIL, decreased from the same initial value of 4,501.92 Teragrams of carbon (TgC) in 1990 to 87.28 TgC and 745.78 TgC in 2050, respectively. As previously stated, emissions under CVL were treated as the baseline, representing the scenario without any project activities, while emissions under RIL accounted for project emissions resulting from implementing project activities. Hence, emissions under RIL were compared to those under CVL to estimate the emission reductions. Findings indicate that total emissions from CVL and RIL over the course of the study were 4590.04 $TgCO_2$ and 3933.37 $TgCO_2$, respectively, per Figure 6. Notably, the switch from CVL to RIL resulted in total emission reductions of 656.67 $TgCO_2$, or 10.94 $TgCO_2$ annually, per Figure 7. We calculated CC under RIL as 7.66 $TgCO_2$ annually, with the aforementioned 30 percent deduction for leakages. Total carbon revenue transitioning from CVL to RIL was estimated at USD 4596.72 million, or USD 76.61 million annually. Another study highlighted the potential for sustainable timber production while reducing emissions by 44 percent through improved methods (Ellis *et al.*, 2019). Appropriate forest management cycles and enhanced logging practices surpass conventional logging practices in terms of carbon stocks and the supply of wood for end uses (Putz *et al.*, 2012).

Similarly, previous studies have provided estimates of total carbon emissions from selective logging in the tropics, including 1870 $TgCO_2$ based on 6 sample blocks in two countries (Putz *et al.*, 2008), 1090 $TgCO_2$

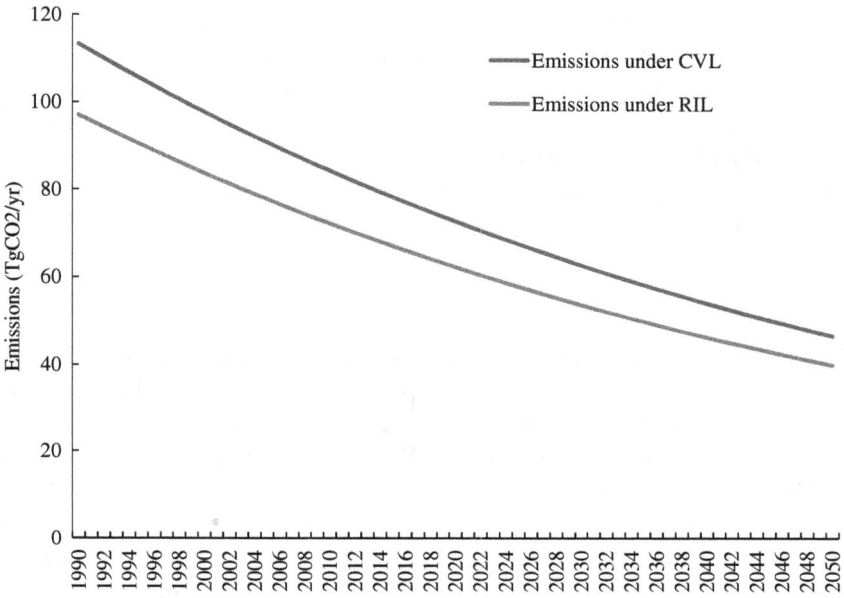

Figure 6. Emissions from CVL and RIL, 1990–2050.

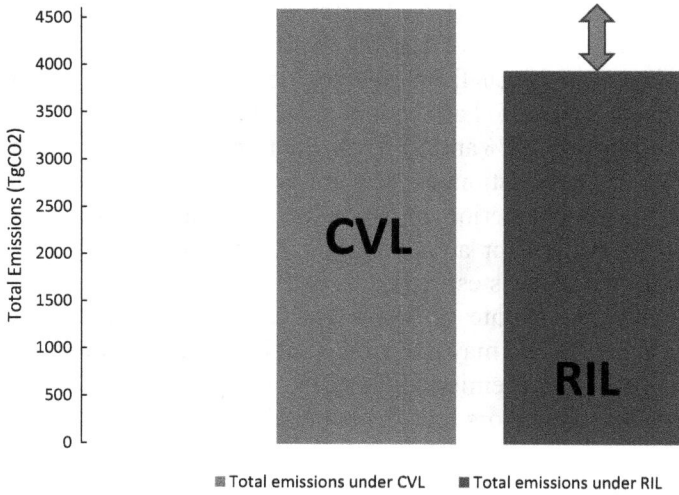

Figure 7. Total emissions from CVL and RIL, 1990–2050.

based on 13 sample blocks in six countries (Pearson *et al.*, 2017), and a model-based estimate of 1923 TgCO$_2$ based on the average of all logging entries (Sasaki *et al.*, 2016). While there is a significant market for logging products, existing CVL techniques leave a negative environmental impact. By reducing damage to surrounding trees and mitigating carbon emissions from decaying trees through RIL, logging-related carbon emissions can be reduced by up to 30 percent (Koester, 2010; ITTO, 2023).

3.4. *Carbon stocks and carbon sequestration in FPF and SPF*

This study demonstrates that the total predicted carbon stocks in both FPF and SPF were 184.90 TgC in 1990, gradually increasing to 594.69 TgC in 2050 with an annual growth of 6.83 TgC. Notably, this represents an annual increase of 3.69 percent, or 6.83 TgC. Specifically, carbon stocks in SPF increased from 101.28 TgC in 1990 to 325.74 TgC in 2050, showing a significant annual growth of 3.74 TgC. Meanwhile, carbon stocks in FPF were estimated to rise from 83.62 TgC in 1990 to 268.94 TgC in 2050, with an annual change of 3.09 TgC. Similarly, Southeast Asian forests continue to act as a source of atmospheric carbon, despite an overall increase in total carbon stock of approximately 20.1 TgC annually between 2000 and 2050, attributed to the ongoing decline of natural forests across Southeast Asia (Kim Phat *et al.*, 2004).

A significant amount of carbon sequestration was observed in both types of plantation forests from 1990 to 2050. Carbon removals from SPF and FPF were 1,928.75 TgCO$_2$ and 1,592.44 TgCO$_2$, respectively, with FPF contributing 45.22% and SPF contributing 54.78% of the total carbon removals. Our study estimated that the potential carbon revenue during the Paris Agreement period of 2020–2030 amounted to approximately USD 2,54.68 million for annual carbon removals of 27.55 TgCO$_2$. The total carbon capture was estimated to be 3521.19 TgCO$_2$, with an annual total carbon removal rate of 58.69 TgCO$_2$. Carbon sequestration from plantation forests alone may contribute 56.37 percent of the total Mekong greenhouse gas (GHG) emissions in 2019, as reported by Climate Watch's climate analysis indicators tool (Climate Watch, 2020).

3.5. *Total carbon capture from PDF and PF*

Total carbon capture, including reductions and removals, from both the PDF and PF, was 3980.86 TgCO$_2$, with an annual capture rate of

66.35 TgCO$_2$. Approximately 88.45 percent of emissions reductions can be attributed to the PF, with the remainder attributed to the PDF. Importantly, this carbon capture has the potential to generate a total carbon revenue of approximately USD 39,808.65 million, assuming the aforementioned rate of USD 10 per ton of CO$_2$. This study finds that PF plays a crucial role in sequestering carbon, contributing substantially to the increment of carbon sinks in the region. Additionally, an estimated USD 3,500.07 million in carbon revenue may be obtained in the aforementioned Paris Agreement period 2020–2030.

The study emphasizes the significant role of switching logging practices from CVL to RIL for carbon capture in the Mekong PDF. Consequently, Mekong PF can function as a carbon sink by adopting RIL, rather than being a carbon source under CVL practices. Implementing RIL can also enhance the economic value of selectively logged forests by capturing carbon through the REDD+ scheme, which can help to prevent these forests from converting into agricultural plantations, avoiding the resultant significant biodiversity loss (Edwards *et al.*, 2012).

According to 2017 estimates of the Food and Agriculture Organization of the UN (FAO), conversion forests contributed to over half of timber production in the Greater Mekong Subregion (GMS) nations. These estimates highlight the significance of timber sourced from conversion forests, even though reliable data on this aspect are limited. Mekong governments have been urged to review their forest management plans, policies, and laws in response to long-standing forest exploitation issues, leading to positive trends that have emerged in past decade. These include increased efforts toward SFM, forest conservation, and afforestation and reforestation programs (Yasmi *et al.*, 2017).

However, transboundary trade between the GMS and neighboring countries contributes to illegal and unsustainable logging practices, exacerbating deforestation challenges. Insufficient measures to ensure SFM and enforce laws have led to persistent issues in this regard. To combat this widespread problem, more resources and initiatives must be allocated toward enhancing law enforcement, combating corruption, and promoting transparency (Costenbader *et al.*, 2015).

Forest plantation programs are crucial in restoring forest cover, providing timber and non-timber forest products, sequestering carbon, and restoring ecosystem services. While plantation forests can help alleviate the pressure on natural forests, they cannot fully replace them. Many nations in the Mekong are implementing REDD+ programs at national and subnational levels alike to increase nature-based carbon capture through forest restoration and SFM practices. These initiatives contribute to carbon emission reduction and offer financial benefits, supporting national and regional development, in addition to improving forest ecosystem services and sustainable timber production.

A study by Guerra-De la Cruz and Galicia (2017) emphasized that forest type, management practices, wood product types, and biomass conversion efficiency all influence the overall impact of forest management on GHGs. Effective forest management can increase carbon sequestration by storing a significant portion of harvested carbon in long-lasting products, preventing its immediate release into the environment.

This study provides valuable insights for regional governments, policymakers, and project developers in the Mekong regarding the importance of SFM and improved logging practices. It highlights the role of these practices in enhancing forest carbon storage, reducing carbon emissions, and ensuring sustainable timber production to achieve sustainable development goals. By incorporating SFM and adopting improved logging practices, stakeholders can optimize the benefits of REDD+ activities and contribute to the region's long-term conservation and utilization of forest resources.

4. Conclusion

Our study aimed to estimate potential nature-based carbon capture in the Mekong by implementing forest management and restoration practices. We specifically focused on the impact of CVL and RIL on carbon emissions and capture in the production forest from 1990 to 2050, corresponding to two 30-year management cycles. Our findings revealed a concerning trend of PDF area decline, with an annual reduction of 0.70 million ha or 0.98 percent over the study period. By contrast, the PF area experienced a modest increase of 0.16 million ha or 3.69 percent per year. This highlights the urgent need for comprehensive measures, including restoration efforts, effective forest management strategies, and robust policies and law enforcement, to halt the decline of natural forests in the Mekong and prevent future timber shortages. By utilizing our model, we observed that CVL produced 9.51 million cubic meters of wood waste, while RIL yielded only 1.27 million cubic meters, representing an average annual reduction of 8.24 million cubic meters. Notably, the transition from CVL to RIL contributed to substantial emissions reductions of 656.67 TgCO$_2$, or 10.94 TgCO$_2$ annually. Significantly, our projections indicated total carbon capture of 3521.19 TgCO$_2$ from the PF.

Our study demonstrates that sustainable timber supply and effective climate change mitigation can be achieved through the adoption of proper logging techniques, specifically RIL, combined with financial subsidies for nature-based carbon capture. Successful implementation of RIL in production forest management supports sustainable development goals and enhances forest ecosystem services. By adopting SFM practices, the Mekong can access carbon-based incentives under the REDD+ framework, contributing to regional development and achieving emission reduction targets.

References

Boltz, F., Holmes, T. P., and Carter, D. R. (2003). Economic and environmental impacts of conventional and reduced-impact logging in Tropical South America: A comparative review. *Forest Policy and Economics*, 5(1), 69–81.

Bösch, M., Elsasser, P., Franz, K., Lorenz, M., Moning, C., Olschewski, R., Rödl, A., Schneider, H., Schröppel, B., and Weller, P. (2018). Forest ecosystem services in rural areas of Germany: Insights from the national TEEB study. *Ecosystem Services*, 31, 77–83.

Bösch, M. (2021). Institutional quality, economic development and illegal logging: A quantitative cross-national analysis. *European Journal of Forest Research*, 140, 1049–1064. https://doi.org/10.1007/s10342-021-01382-z.

Brown, S., Casarim, F. M., Grimland, S. K., and Pearson, T. (2011). *Carbon Impacts from Selective Logging of Forests in Berau, East Kalimantan, Indonesia*. Final Report to The Nature Conservancy.

Butarbutar, T., Soedirman, S., Neupane, P. R., and Köhl, M. (2019). Carbon recovery following selective logging in tropical rainforests in Kalimantan, Indonesia. *Forest Ecosystems*, 6(1), 1–14.

Calvo-Alvarado, J. C., Arias, D., and Richter, D. D. (2007). Early growth performance of native and introduced fast growing tree species in wet to sub-humid climates of the Southern region of Costa Rica. *Forest Ecology and Management*, 242(2–3), 227–235.

Climate Watch. (2020). *GHG Emissions*. Washington, DC: World Resources Institute. Available at: https://www.climatewatchdata.org/ghg-emissions.

Costenbader, J. *et al.* (2015). Drivers of forest change in the Greater Mekong Subregion —Regional report. USAID-LEAF and FAO.

Dar, J. A. and Sundarapandian, S. (2015). Variation of biomass and carbon pools with forest type in temperate forests of Kashmir Himalaya, India. *Environmental Monitoring and Assessment, 187*(2). https://doi.org/10.1007/s10661-015-4299-7.

Edwards, D. P., Woodcock, P., Edwards, F. A., Larsen, T. H., Hsu, W. W., Benedick, S., and Wilcove, D. S. (2012). Reduced-impact logging and biodiversity conservation: A case study from Borneo. *Ecological Applications, 22*(2), 561–571. https://doi.org/10.1890/11-1362.1.

Ellis, P. W., Gopalakrishna, T., Goodman, R. C., Putz, F. E., Roopsind, A., Umunay, P. M., Zalman, J., Ellis, E. A., Mo, K., Gregoire, T. G., and Griscom, B. W. (2019). Reduced-impact logging for climate change mitigation (RIL-C) can halve selective logging emissions from tropical forests. *Forest Ecology and Management, 438*, 255–266. https://doi.org/10.1016/j.foreco.2019.02.004.

FAO. (2010). *Global Forest Resources Assessment.* Main report. FAO Forestry Paper 163.

FAO. (2011). *Forests and Forestry in the Greater Mekong Subregion to 2020.* Accessed September 2022.

FAO. (2015). *Land Use. Forest Area. Cambodia, Laos, Myanmar, Thailand & Vietnam (2000–2015).* Accessed September 2022.

FAO. (2000). *Global Forest Resources Assessment (main report).* FAO forestry paper, 140, Rome.

FAO. (2001). *Mean Annual Volume Increment of Selected Industrial Forests Plantation Species.* Forets plantation thematic papers, Working Paper 1.

FAO. (2020). *Global Forest Resources Assessment 2020: Main Report.* Rome.

FAO. (2016). *State of the World's Forests 2016; Forests and Agriculture, Land-Use Challenges and Opportunities.* Rome: Food and Agriculture Organization.

Guerra-De la Cruz, V. and Galicia, L. (2017). Tropical and highland temperate forest plantations in Mexico: Pathways for climate change mitigation and ecosystem services delivery. *Forests, 8*(12), 489. https://doi.org/10.3390/f8120489.

Holmes, T. P., Blate, G. M., Zweede, J. C., Pereira, R., Barreto, P., Boltz, F., and Bauch, R. (2000). *Financial Costs and Benefits of Reduced Impact Logging Relative to Conventional Logging in the Eastern Amazon.* Washington, DC: Tropical Forest Foundation.

Huy, L. Q. (2004). *Fast-growing Species Plantations–Myths and Realities and Their Effect on Species Diversity.* Himachal Pradesh, India: Dr. YS Parmar University of Horticulture and Forestry.

IPCC. (2006). *IPCC Guidelines for National Greenhouse Gas Inventories.* Hayama: Institute for Global Environmental Strategies.

ITTO. (2005). Revised ITTO criteria and indicators for the sustainable management of tropical forests including reporting format. ITTO Policy Development Series No. 15. ITTO, Yokohama, Japan.

Khai, T. C., Mizoue, N., and Ota, T. (2020). Post-harvest stand dynamics over five years in selectively logged production forests in Bago, Myanmar. *Forests*, *11*(2), 195. https://doi.org/10.3390/f11020195.

Kim Phat, N., Knorr, W., and Kim, S. (2004). Appropriate measures for conservation of terrestrial carbon stocks — Analysis of trends of forest management in Southeast Asia. *Forest Ecology and Management*, *191*(1–3), 283–299. https://doi.org/10.1016/j.foreco.2003.12.019.

Kinoshita, T., Inoue, K., Iwao, K., Kagemoto, H., and Yamagata, Y. (2009). A spatial evaluation of forest biomass usage using GIS. *Applied Energy*, *86*(1), 1–8. https://doi.org/10.1016/j.apenergy.2008.03.017.

Koester, S. (2010). Reduced impact logging. *Forest Ecology and Management*, *256*, 1427–1433.

ITTO International Tropical Timber Organization. https://www.itto.int/sustainable_forest_management/logging/.

Lowe, A. J., Dormontt, E. E., Bowie, M. J., Degen, B., Gardner, S., Thomas, D., Clarke, C., Rimbawanto, A., Wiedenhoeft, A., Yin, Y., and Sasaki, N. (2016). Opportunities for improved transparency in the timber trade through scientific verification. *BioScience*, *66*(11), 990–998. https://doi.org/10.1093/biosci/biw129.

MacDicken, K. G., Sola, P., Hall, J. E., Sabogal, C., Tadoum, M. and de Wasseige, C. (2015b). Global progress toward sustainable forest management. *Forest Ecology and Management*, *352*, 47–56.

Matangaran, J. R., Putra, E. I., Diatin, I., Mujahid, M., and Adlan, Q. (2019). Residual stand damage from selective logging of tropical forests: A comparative case study in central Kalimantan and West Sumatra, Indonesia. *Global Ecology and Conservation*, *19*, e00688.

Nizami, S. M., Yiping, Z., Liqing, S., Zhao, W., and Zhang, X. (2014). Managing carbon sinks in rubber (*Hevea brasilensis*) plantation by changing rotation length in SW China. *PLoS ONE*, *9*(12), e115234. https://doi.org/10.1371/journal.pone.0115234.

Pandey, D. and Brown, C. (2000). Teak: A Global Overview. *UNASYLVA-FAO*, *51*, 3–13.

Pearson, T. R. H., Brown, S., Murray, L., and Sidman, G. (2017). Greenhouse gas emissions from tropical forest degradation: An underestimated source. *Carbon Balance and Management*, *12*(1). https://doi.org/10.1186/s13021-017-0072-2.

Piponiot, C., Derroire, G., Descroix, L., Mazzei, L., Rutishauser, E., Sist, P., and Hérault, B. (2018). Assessing timber volume recovery after disturbance in tropical forests — A new modelling framework. *Ecological Modelling*, *384*, 353–369.

Piponiot, C., Rödig, E., Putz, F. E., Rutishauser, E., Sist, P., Ascarrunz, N., Blanc, L., Derroire, G., Descroix, L., Guedes, M. C., Coronado, E. H., Huth, A., Kanashiro, M., Licona, J. C., Mazzei, L., d'Oliveira, M. V. N., Peña-Claros, M., Rodney, K., Shenkin, A., and Hérault, B. (2019a). Can timber provision from Amazonian production forests be sustainable? *Environmental Research Letters, 14*(6), 064014. https://doi.org/10.1088/1748-9326/ab195e.

Piponiot, C., Rutishauser, E., Derroire, G., Putz, F. E., Sist, P., West, T. A. P., Descroix, L., Guedes, M. C., Coronado, E. N. H., Kanashiro, M., Mazzei, L., d'Oliveira, M. V. N., Peña-Claros, M., Rodney, K., Ruschel, A. R., Souza, C. R. D., Vidal, E., Wortel, V., and Hérault, B. (2019b). Optimal strategies for ecosystem services provision in Amazonian production forests. *Environmental Research Letters, 14*(12), 124090. https://doi.org/10.1088/1748-9326/ab5eb1.

Plath, M., Mody, K., Potvin, C., and Dorn, S. (2011). Establishment of native tropical timber trees in monoculture and mixed-species plantations: Small-scale effects on tree performance and insect herbivory. *Forest Ecology and Management, 261*(3), 741–750.

Player, B. and Tabar, J. (2021, October 4). *Ideas Capturing Carbon: How Nature-Based Solutions Help Achieve Net Zero Goals.* https://www.stantec.com/. Retrieved September 25, 2022, from https://www.stantec.com/en/ideas/capturing-carbon-why-nature-based-solutions-are-the-tool-of-choice-to-achieve-net-zero-goals.

Putz, F., Sist, P., Fredericksen, T., and Dykstra, D. (2008). Reduced-impact logging: Challenges and opportunities. *Forest Ecology and Management, 256*(7), 1427–1433. https://doi.org/10.1016/j.foreco.2008.03.036.

Putz, F. E., Zuidema, P. A., Synnott, T., Peña-Claros, M., Pinard, M. A., Sheil, D., Vanclay, J. K., Sist, P., Gourlet-Fleury, S., Griscom, B. W., Palmer, J., and Zagt, R. (2012). Sustaining conservation values in selectively logged tropical forests: the attained and the attainable. *Conservation Letters, 5*(4), 296–303. https://doi.org/10.1111/j.1755-263x.2012.00242.x.

RECOFTC. (2018). *Overcoming Threats to the Mekong's Forests and People.* Bangkok: RECOFTC. https://www.recoftc.org/special-report/forest-governance-mekong.

Sari, D. R. and Ariyanto. (2018). The potential of woody waste biomass from the logging activity at the natural forest of Berau District, East Kalimantan. *IOP Conference Series: Earth and Environmental Science, 144*, 012061. https://doi.org/10.1088/1755-1315/144/1/012061.

Sasaki, N. (2021). Timber production and carbon emission reductions through improved forest management and substitution of fossil fuels with wood biomass. *Resources, Conservation and Recycling, 173*, 105737. https://doi.org/10.1016/j.resconrec.2021.105737.

Sasaki, N., Asner, G. P., Pan, Y., Knorr, W., Durst, P. B., Ma, H. O., Abe, I., Lowe, A. J., Koh, L. P., and Putz, F. E. (2016). Sustainable management of tropical forests can reduce carbon emissions and stabilize timber production. *Frontiers in Environmental Science, 4.* https://doi.org/10.3389/fenvs.2016.00050.

Sasaki, N., Chheng, K., and Ty, S. (2012). Managing production forests for timber production and carbon emission reductions under the REDD+ scheme.

Environmental Science & Policy, 23, 35–44. doi: 10.1016/j.envsci.2012. 06.009.

Sasaki, N., Myint, Y. Y., Abe, I., and Venkatappa, M. (2021). Predicting carbon emissions, emissions reductions, and carbon removal due to deforestation and plantation forests in Southeast Asia. *Journal of Cleaner Production, 312*, 127728. https://doi.org/10.1016/j.jclepro.2021.127728.

Sloan, S. and Sayer, J. A. (2015). Forest Resources Assessment of 2015 shows positive global trends but forest loss and degradation persist in poor tropical countries. *Forest Ecology and Management, 352*, 134–145.

Smith, P., Clark, H., Dong, H., Elsiddig, E. A., Haberl, H., Harper, R., ... and Tubiello, F. (2014). *Agriculture, Forestry and Other Land Use (AFOLU).* Cambridge: Cambridge University Press.

Stibig, H. J., Stolle, F., Dennis, R., and Feldkötter, C. (2007). Forest cover change in Southeast Asia — The regional pattern. *JRC Scientific and Technical Reports, EUR, 22896.*

Taylor, R. (2021). *What COP26 Means for Forests and the Climate.* World Resources Institute. https://www.wri.org/insights/what-cop26-means-forests-climate.

Ty, S., Sasaki, N., Ahmad, A., and Ahmad, Z. (2011). REDD development in Cambodia — Potential carbon emission reductions in a REDD project. *FORMATH, 10*(0), 1–23. https://doi.org/10.15684/formath.10.1.

UNFCCC. (2011). *Indicative Simplified Baseline And Monitoring Methodologies For Selected Small-Scale CDM Project Activity Categories: General Guidance On Leakage In Biomass Project Activities* (Version 03). http://cdm. unfccc.int/Reference/Guidclarif/ssc/methSSC_guid04.pdf.

UNFCCC. (2008, March). Report of the Conference of the Parties on Its Thirteenth Session, held in Bali from 3 to 15 December 2007. In Addendum. Part Two: Action taken by the Conference of the Parties at its thirteenth session Decisions adopted by the Conference of the Parties.

UNFCCC. (2010). Report of the Conference of the Parties on Its Sixteenth Session, Held in Cancun from 29 November to 10 December 2010. Addendum; Part Two: Action Taken by the Conference of the Parties at Its Sixteenth Session. https://unfccc.int/documents/6527.

Vasco, C., Torres, B., Pacheco, P., and Griess, V. (2017). The socioeconomic determinants of legal and illegal smallholder logging: Evidence from the Ecuadorian Amazon. *Forest Policy and Economics, 78*, 133–140.

World Bank. (2020). *State and Trends of Carbon Pricing 2020.* World Bank, Washington, DC: Tech. Rep.

World Bank. (2022, May 17). *What You Need to Know About Nature-Based Solutions to Climate Change.* World Bank. Retrieved September 26, 2022, from https://www.worldbank.org/en/news/feature/2022/05/19/what-you-need-to-know-about-nature-based-solutions-to-climate-change.

WWF. (2018). *Pulse of the Forest: The State of the Greater Mekong's Forests and the Everyday People Working to Protect Them.* Vaud: World Wildlife Fund.

Yasmi, Y., Durst. P., Haq, R. U., and Broadhead, J. (2017). *Forest Change in the Greater Mekong Subregion (GMS): An Overview of Negative and Positive Drivers.* Bangkok: FAO.

Chapter 6

Development of Graphene as Advanced Functional Material for Carbon Capture Technology: A Membrane Technology Approach for CCUS

**Andri Hardiansyah[*,§], Ni Luh Wulan Septiani[*,¶],
Alan Ray Farandy[†,‖], Ruth Meilianna[‡,**],
and Rizki Febrian[*,††]**

*Research Center for Advanced Materials,
National Research and Innovation Agency (BRIN),
Tangerang Selatan, Banten, 15314, Indonesia*

†*Research Center for Behavior and Circular Economics,
National Research and Innovation Agency (BRIN),
Jakarta Selatan, 12710, Indonesia*

‡*Research Center for Population,
National Research and Innovation Agency (BRIN),
Jakarta Selatan, 12710, Indonesia*

§*andri.hardiansyah@brin.go.id*

¶*nilu010@brin.go.id*

‖*alan002@brin.go.id*

**ruth002@brin.go.id*

††*rizki.febrian@brin.go.id*

Abstract

While the Association of Southeast Asian Nations (ASEAN) is a leading contributor of greenhouse gases (GHGs) due to its dependency on fossil fuels, a sudden transition to renewable energy would be cost-intensive and difficult due to a lack of infrastructure. Carbon Capture, Utilization, and Storage (CCUS) can be implemented to combat carbon dioxide (CO_2) emissions. Membrane technology offers to lower CCUS costs and is made more viable by incorporating advanced materials, such as graphene-based materials (GBM) and other 2D substances, to improve carbonaceous-based gas separation and selectivity. Herein, we develop the GBM membrane, evaluating its membrane structure through material characterization and its sensing performance on gaseous compounds. We also investigate the prospects, utilization, and challenges of membrane technology in CCUS, particularly in ASEAN and Indonesia. Scanning electron microscopy (SEM) examinations show that the GBM has a porous layered structure and performs its sensing character of carbonaceous-based gases. This study shows that membrane technology could be implemented and would fulfill ASEAN demands as well as meet carbon neutrality targets with government policy and infrastructure support.

Keywords: Carbon capture, graphene, membrane, technology.

1. Introduction

The trend of atmospheric carbon dioxide (CO_2) concentration has seen an increase over the past century. Further increases are anticipated as a result of the rapidly rising use of fossil fuels to meet the rising energy needs of an expanding population. Efforts must be made to reduce atmospheric CO_2 because it affects the global climate. (Al-Mamoori *et al.*, 2017; Gizer *et al.*, 2022; Yu *et al.*, 2008). Everyone agrees that the most efficient and long-term way of stabilizing atmospheric CO_2 concentrations would be developing and implementing a diversified portfolio of sophisticated energy technologies. Carbon capture, utilization, and storage (CCUS) is a strategy for decreasing anthropogenic energy-related CO_2 emissions through greater energy efficiency and increased renewable usage. Various strategies, including pre-combustion, post-combustion, and oxy-fuel combustion, and various technologies, such as absorption, adsorption, membrane separation, and cryogenic distillation (Soujanya, 2020), are currently being investigated for capturing CO_2 from fossil fuel-burning power plants and industrial sources.

Developing materials and technologies for affordable and energy-efficient CO_2 capture is crucial to reducing the consequences of global warming. Researchers working in this area are focusing on physicochemical-based absorption, membrane-based separation, ionic liquids, and electro-chemical and enzymatic conversion. Membrane-based CO_2-capturing technologies have several advantages, including ecological friendliness, ease of design and operation, and appropriate thermal and mechanical stability, while effectively reducing emissions based on their materials components, structure, morphology, and separation techniques. Membranes are now classed as inorganic membranes, polymeric membranes, or assisted transport membranes (Shah *et al.*, 2021).

Currently, graphene has attracted great interest for its novelty. It is a polymorph of carbon that is flat and two-dimensional with a hexagonal lattice structure. Graphene oxide (GO) has such potential as large surface area, mechanical stability, and electrical and optical properties. It is developed by oxidizing graphite to form graphite oxide, which is contained with oxygen functional groups including epoxy, carbonyl, carboxyl, and hydroxyl groups, and is then exfoliated to form GO. Membranes based on GO are attracting considerable interest in CCUS applications.

This study aims to evaluate CCUS circumstances and policy, particularly in ASEAN and Indonesia, and the implementation of technologies to overcome its drawbacks, with emphasis on membrane technology, concluding with policy recommendations. We use an experimental chemical approach to fabricate a graphene-based material membrane for carbon capture technology to obtain gas capture and sensors with optimal performance. We also evaluate the current membrane-based technological approach as a potential solution for CCUS.

2. Literature Review

2.1. *Current situation: Worldwide and Indonesia*

Cost and efficiency calculations by country must consider the value chains and technologies used to avoid bias. Several technologies are used in CCUS, which are divided into engineered technologies for carbon capture, i.e., fossil fuels with CCS, direct air carbon capture and storage

(DACCS), biomass energy with carbon capture and storage (BECCS), and carbon storage technologies, i.e., storage into aquifers, enhanced oil recovery (EOR), and carbon use technologies. BECCS allows the removal of atmospheric CO_2 by plants. The resulting biomass is burned, and the carbon stored therein is recovered. With DACCS, CO_2 is captured directly from water. In both cases, captured CO_2 is compressed and injected into layers of porous rock a kilometer or more underground, under an impermeable rock that will retain it for tens of thousands to millions of years. While DACCS is estimated to be twice the cost of BECCS, DACCS is able to remove large amounts of CO_2 without the demands on natural systems required by biomass growth (UNECE, 2021).

EOR improves oil and gas recovery by injecting CO_2 under pressure as a secondary propulsion mechanism to drive carbon remaining in oil and gas reservoirs. CO_2 can also be injected with water. Sources are based on local availability, with the cheapest and most abundant available from natural sources. CO_2 EOR must become economically competitive versus drilling new wells and other techniques (UNECE, 2021). This depends on the suitability of a given reservoir, the required payback period due to relatively high capital costs, local CO_2 costs, and the availability of technical resources.

Cooperation between member countries will provide the most effective and efficient mitigation strategies for subsurface CO_2 storage and sequestration. Collaboration on existing shared regional CO_2 transport and storage infrastructure is required if CCUS is to be deployed on a large enough scale to contribute significantly to carbon neutrality (UNECE, 2021).

Each country has a different number of projects. Table 1 shows one project from each surveyed country having the highest carbon capture of all projects owned by the country concerned, with the total cost per project. The cost for the UK is USD 20 million and the Iceland project is worth USD 500 million, owing to Iceland's higher project efficiency. Countries such as Sweden have lower project and efficiency values than other countries.

China, Belgium, and Canada use EOR, with values of USD 13 million, USD 11 million, and USD 1.9 billion, respectively, albeit with relatively lower efficiency scores than other countries at 0.34, 0.2, and 0.21, respectively. Industrial capture projects are employed by Norway, the United States, Germany, Croatia, the Netherlands, Australia, and the United Arab Emirates, with project expenditures varying across these seven countries, ranging from as low as $6 million in Iceland to as high as $1 billion in Australia. Efficiency values range from a high of 0.85 in the US to lows in Australia and Norway, showing that project value does

Table 1. Surveyed countries.

Countries	Project name	Project type	Cost (in USD)	Carbon capture/year	Efficiency
France	dmx demonstration in dunkirk	CO_2 capture and storage	$20 million	3	0.1
UK	HYNET	CO_2 capture and storage	$13 million	1	0.34
China	Jilin Oil Field CO_2-EOR	Industrial capture, EOR	$11 million	0.6	0.2
Belgium	leilac	Industrial capture, EOR	$20 million	1	0.24
Norway	Sleipner CO_2 Storage	Industrial capture	$63 million	1	0.34
Qatar	Qatar LNG CCS	Power and capture	$200 million	2.1	0.71
Greece	ENERGEAN CARBON STORAGE	CO_2 Storage	$500 million	2.5	0.8
Iceland	Silverstone	CO_2 capture and storage	$6 million	0.034	0.85
USA	Illinois Industrial Carbon Capture and Storage (ICCS)	Industrial capture	$208 million	1	0.34
Germany	leilac 2	Industrial capture	$26 million	0.1	0.34
South Africa	Pilot Carbon Storage Project (PCSP)	N/A	$27 million	0.03	0.1
Sweden	VETENFALL UPPSALA	CO_2 capture	$318 million	0.2	0.68
Croatia	Icord	industrial capture	$52877 million	1	0.31
The Netherlands	Porthos (Port of Rotterdam)	Industrial capture	$2 billion	5	0.31
Italy	CCS Ravenna Hub	CO_2 capture, transportation and storage	$1 billion	2.02	0.24
Australia	Gorgon Carbon Dioxide Injection	Industrial capture	$1 billion	3.7	0.28
Poland	Poland EU CCS Interconnector	CO_2 transport and storage	$1.9 billion	2.7	0.21
Canada	quest	Industrial capture, EOR	$1.4 billion	1	0.97
United Arab Emirates	Abu Dhabi CCS (Phase 1: Emirates Steel C industries)	Industrial capture	$15 billion	0.8	0.1

Source: UNECE (2021) (modified).

not necessarily correlate with efficiency value. Italy has a CO_2 capture, transport and storage project, and Poland has a CO_2 transport and storage project, the latter having a project value of USD 1 billion, with an efficiency value of only 0.24.

Indonesia has a first nationally determined contribution (NDC), which is a document submitted to the UNFCCC to express Indonesia's commitment to low emissions and climate-resilient development and to join the global effort to prevent a 1.5°C rise in global average temperature above pre-industrial levels. It commits Indonesia to reduce greenhouse gas emissions (GHGs) by 31.89 percent through domestic efforts and by 43.2 percent through international collaboration by 2030, relative to the baseline business-as-usual trajectory. The goal of the CCUS development in Indonesia is to reduce GHGs for oil and gas fields and electric power plant emissions. The Gundih Carbon Capture and Storage Pilot Project is the first CCS project in South and Southeast Asia (Sule & Widiyantoro, 2018).

The Bandung Institute of Technology (ITB), Pertamina, Ministry of Energy and Mineral Resources, Kyoto University, JST, and JICA have been researching the implementation of CCS/CCUS in the Gundih Project since 2012, and the Gundih field has been producing gas since the end of 2013 at a CO_2 concentration of 21 percent. The Central Processing Plant (CPP) separates approximately 15% of the overall product as nearly pure CO_2 or 800 tonnes daily. The Kedung Tuban (KTB), Randu Blatung (RBT), and Kedung Lusi (KDL) structures located within the Gundih field produce gas, approximately 60 MMSCFD to date.

Currently, 15 CCS/CCUS projects in Indonesia are in the study phase, and it is expected that all will be operational by 2030. The Tangguh EGR/CCUS project, managed by BP Berau Ltd., is targeted to come online in 2026 with a CO_2 potential of 25–32 million tons over 10 years. Potential CO_2 storage in Indonesia is of the order of two gigatonnes of CO_2, particularly in Sulawesi, which cooperates with Japan, Kalimantan, where Pertamina cooperates with ExxonMobil, and the Sukowati Field, where Pertamina collaborates with Japan.

The Ministry of Energy and Mineral Resources (MEMR) has prepared a ministerial regulation draft on CCS/CCUS in oil and gas working areas, emphasizing technical aspects based on applicable regulation, referenced standards, good engineering practices, and site characteristics, along with CCS/CCUS monetization opportunities. The draft has been shared with stakeholders for comment, including related Indonesian ministries, international institutions, public or private institutions, research centers, and companies from the UK, Australia, Japan, the EU, and the US. This regulation is significant because CCUS plays a significant role in lowering emissions from the natural gas value chain, especially CO_2 removal from methane to meet regulations for natural gas or LNG. This procedure produces a CO_2 stream with greater than 98 percent concentration, making natural gas processing one of the least expensive CCUS applications.

GHG being a global problem necessitates a global solution. The benefits of reducing greenhouse gases occur to individuals, institutions, and countries alike as well. Cooperation between nations and regions is one strategy for solving these global issues, having the potential to overcome such challenges as the many actors with different perspectives regarding costs and benefits of collective action, sources of emissions, variation of climate impacts, and mitigation costs (Stavins *et al.*, 2014).

Japan's participation in the CCUS development in Indonesia is crucial. At least six of the 15 CCUS development projects in Indonesia involve the Japanese public or private sectors (MEMR, 2022). First, Japan Gasoline Co (JGC), J-Power, and Japan Nus Co (Janus) participated in the first CCUS project in Indonesia, the Gundih site, which was also supported by the Japanese Ministry of Economy, Trade, and Industry (METI). This project is currently in phase 2 research to mitigate uncertainties and risks, with an onstream target date of 2026 and a potential storage capacity of 3 million tCO_2 over 10 years. Second, in the Sukowati project, Pertamina works together with Lemigas and Japan Petroleum Exploration (Japex), again with METI support. Pilot testing is planned for 2026–2027 and go fully online in 2031, with a 7–14 million tCO_2 reduction over 15 years. Third is the Abadi CCUS project in the south of East Nusa Tenggara, which is being developed by Inpex Masela and could store 70 million tons of native CO_2 by 2055. Fourth, Pertamina, PT. Panca Amara Utama, ITB, the Japan Organization for Metals and Energy Security (Jogmec), and Mitsubishi launched the Joint Study Blue Ammonia & CCS in Central Sulawesi, which is anticipated to undergo preliminary feasibility testing in July 2023, with 19 million tCO_2 reduction. Other projects include a study of CCUS for Coal to DME in South Sumatra and the Central Sumatra Basin CCS/CCUS Regional Hubs project that involves Pertamina, Chiyoda Corporation, and Mitsui.

Japan participates in the CCUS project finance as well as in the collaboration on CCUS development. Funding can be in the form of a grant, such as that which the Gundih project received in 2011 from The Science and Technology Research Partnership for Sustainable Development (SATREPS) in response to a proposal submitted to JICA and JST in October 2010. The SATREPS project, which began in September 2012 and ended in September 2017, included greenhouse gas removal (GGR) for site selection and monitoring. Other financing options include the joint crediting mechanism (JCM) scheme, an initiative of the Japanese government that provides incentives for Japanese companies to engage in low-carbon development projects in Indonesia. The Japanese government would then be able to claim a share of reduced GHGs from investment projects in Indonesia as emission reductions for Japan. The implemented project is the Gundih project, a joint study with Kanso Technos in 2015 using the JCM mechanism (Sule & Widiyantoro, 2018).

Indonesia still requires international collaboration to accelerate its transition to carbon neutrality. The Indonesian Ministry of Energy and Mineral Resources underlined the necessity for international cooperation and collaboration in the following areas (MEMR, 2022): (1) mapping of potential CO_2 storage capacity in saline aquifers, (2) CO_2 utilization from oil and gas field to produce methanol, (3) feasibility studies of blue hydrogen development, (4) studies of business models and MRVs for CCS hubs, and (5) capacity building for increased CCS/CCUS implementation in seminars, training, or courses.

2.2. CCUS membranes

A crucial step in CCUS is carbon capture, as it directly affects GHGs by absorbing CO_2 from both the atmosphere and combustion flue gas. This is also the most cost-intensive process in the CCUS supply chain (Hasan *et al.*, 2015). Membrane technology might help, with its low energy demand and operational costs, small footprint, and the ease with which it can be scaled up and incorporated into existing technologies (Kárászová *et al.*, 2020). According to Sandra E. Kentish, CCS has a significant role in climate change solutions and the hydrogen economy, as well as manufacturing steel, cement, and chemicals (Kentish, 2019). Various fabrication methods are also available (Shi *et al.*, 2022).

Membrane function in CCUS is similar to a filter that removes permeated gas and retains retentate gases. Its performance as a separator depends on conditions, such as pore size and sizes of gas molecules, the affinity of the gas to the membrane, and the molecular weight of the gas.

In general, to maximize performance, the feeding gas is set to high pressure, while the permeated gas is connected to an area of low atmospheric pressure, even vacuum. However, these conditions can damage membranes, which have thicknesses ranging from hundreds of nanometers to several micrometers, resulting in cracks that may shorten the life of the membrane. Thus, membranes are typically coated atop substrates to improve mechanical properties (Favre, 2022).

Several commercial polymer membranes, including PRISM™, Polaris™, and Polyactive™, were seen as candidates and accordingly tested at fossil fuel-burning power plants (Kárászová *et al.*, 2020). Although their performance was comparable to, if not better than, monoethanolamine (MEA) absorption, they lacked stability and longevity due to exposure to acid gases (NO_x, SO_2) during use. Membranes' separation performance depends on selectivity and gas permeance optimization due to Robeson's upper bound (He, 2018a).

Several experimental membranes were also studied, including microporous organic polymers (MOPs), fixed-site-carrier membranes (FSCs), mixed matrix membranes (MMMs), and carbon molecular sieve membranes (CMSMs) (He, 2018a). MMMs offer better permeability and separation performance as well as mechanical strength compared to common polymer membranes by combining a polymer matrix with various inorganic fillers (Huang *et al.*, 2018). Incorporating microporous filler may enhance selectivity through molecular sieving or surface flow transport (He, 2018a). One issue that emerged from the MMS studies is defect formation due to the polymer matrix and inorganic filler being incompatible, with said defects detracting from the mechanical and separation performance (Huang *et al.*, 2018).

As mentioned earlier, membranes have several advantages over commercial CCUS technologies, starting with lower costs, deriving from simplicity of construction and maintenance, the latter involving only membrane replacement. Second is chemical stability, meaning that membranes react hardly at all with filtered gases, resulting in long life spans. Third is efficiency, in that membranes can be easily integrated with other systems and packaged into small portable modules. Membranes can also reduce energy consumption by up to 7.5% more than amine absorption technology with yields of more than 99% (Hou *et al.*, 2022).

Performance depends on materials, which must satisfy several criteria including high permeability, allowing high mass transfer flux; high selectivity, determined by permeability ratios of different gases; being free of defects; the stability, chemical, mechanical, and thermal; and low fouling, where unwanted gas adsorption is prevented (He, 2018b). The most popular membrane material to date is polymer because of its ease of fabrication, low cost, and relatively high specific surface area (Chen *et al.*, 2022a). Several polymers reported used in CCUS are rubbery (poly(dimethylsiloxane) and ethylene oxide/propylene oxide–amide copolymers) and glassy polymers (cellulose acetate, tetra bromo bisphenol A (e.g., Matrimid, P84), polycarbonate, polyimides, poly (2,6 dimethyl phenylene oxide), and polysulfone (Hou *et al.*, 2022). Several other polymers have been developed for improved CO_2 capture performance, some of which are described in the following sections.

2.2.1. *Polymer-based membranes*

In thin film composite (TFC) technology, the polymer is coated atop a porous support layer. Two types of TFC have been developed: the Loeb-Sourirajan membrane, which consists of one type of polymer, and the multilayer composite membrane, which consists of several types of polymers. Typically, 50 g of polymer is required per 1 m^2 of membrane in a single membrane, whereas 1–2 g of polymer per 1 m^2 of membrane is required in a multilayer composite membrane (Hou *et al.*, 2022). Gas separation that occurs mainly in non-porous membranes is solution diffusion, with the driving force coming from the difference in partial pressure. Gas separation occurs when CO_2 gas is adsorbed on the surface of the membrane, dissolves into the bulk membrane, diffuses toward the permeated side, and is desorbed back into CO_2. It dates back to 1982, especially the CO_2/N_2 variety, in which cellulose nitrate was combined with polyethylene glycol (PEG), and the optimal ratio between the two may increase CO_2 selectivity (Kawakami *et al.*, 1982). Subsequent studies reported that polyimides incorporated 6FDA had good physical properties and were able to form well membranes that were sufficiently permeable and selective to CO_2 (Niwa *et al.*, 2000). Another combination is PEO segmented copolymers with polyimides (Powell & Qiao, 2006), which are also highly permeable and selective to CO_2. It is believed that this is due to CO_2 being highly soluble in PEO. Composite membranes can also be fabricated by the mixing of inorganic materials with a polymer matrix, which can increase the chemical, mechanical, and thermal matrix stability. Carbon material, specifically the GO, is used as an additive in the mix of polymer polydimethylsiloxane (PDMS) and polyimide, which produces 40–90 GPU CO_2 permeance with 30–44 CO_2/N_2 selectivity (Liang *et al.*, 2017). Metal organic frameworks (MOFs) are inorganic materials with easily adjustable microporosity that supports sieving. Although low chemical stability and relatively complex synthesis limit their use in membranes, they remain in development. MOF-74-Ni and NH_2-UiO-66 have been used as PIM-1 additives (Liu *et al.*, 2020a). CO_2 permeance and selectivity increased from 4320 GPU and 19 to 4660–7460 and 26–33 with MOFs. Combination of polyimide and PIM-1 with HKUST-1 was also reported to increase CO_2 permeance and selectivity (Elsaidi *et al.*, 2021).

2.2.2. *MOF-based thin films*

As mentioned earlier, MOFs are popular in CO_2 adsorption as their porosity can be adjusted to molecular size, allowing sieving in CO_2 capture or gas separation (Venna & Carreon, 2015). Low yields of MOF-based membranes arise from the difficulty in fulfilling the requirements that they are free of defects and moisture-resistant. Chernikova *et al.* applied liquid-phase epitaxy to grow fluorinated MOFs free of defects, SIFSIX-3-M (M = Ni or Cu) on porous alumina substrate as a membrane for CO_2 capture (Chernikova *et al.*, 2020). With 30:70 CO_2/H_2 and 50:50 CO_2/CH_4 mixtures, they found a separation factor for CO_2 to CH_4 of 4 in single gas testing, while assessment of SIFSIX-3-Ni in real separation showed selectivity for CO_2 versus H_2 or CH_4 of 20.4 and 9.9, respectively. Zeolite imidazole framework-based MOFs also have strong CO_2 separation. While ZIF-8 was reported to have a pore aperture of 0.34 nm that is ideal for CO_2/N_2 separation, its use is limited by its flexible lattice which lacks selectivity. Babu *et al.* proposed rapid heat treatment to solve the problem. Heating ZIF-8 for a few seconds at 360°C (Babu *et al.*, 2019) stiffens the lattice, increasing selectivity, where CO_2/CH_4, CO_2/N_2, and H_2/CH_4 selectivity is found to be more than 30, 30, and 175, respectively. Another study reports on modifying UiO-66 with polydopamine (PDA) for improved selectivity (Wu *et al.*, 2019). PDA improves CO_2 capture capability by blocking off pinholes and grain boundaries that restrict CO_2 transport. PDA modified UiO-66 selectivity rose to 13.6 (H_2/N_2), 7.7 (H_2/CH_4), 47.4 (CO_2/N_2), and 26.9 (CO_2/CH_4) from 10.8 (H_2/N_2), 6.9 (H_2/CH_4), 26.5 (CO_2/N_2), and 17.0 (CO_2/CH_4) for unmodified UiO-66.

2.2.3. *Graphene-based inorganic membranes*

Development of carbon-based materials for membranes is attracting interest from science and industry alike for their advantages of low cost, high surface area, potential pore structure modification, and surface functionalization. Graphene, a flat, two-dimensional carbon polymorph with sp^2 hexagonal bonding configuration, is drawing particular attention (Park, 2014). It is popular across a wide range of applications due to its unique physicochemical properties, including high chemical and thermal stability, strong mechanical properties, high conductivity, flexible functionalization, and stability in acid or base states. Graphene-based carbon can be found in three forms according to their surface groups: graphene, GO, and reduced graphene oxide (rGO). GO has potential properties, including large surface area, mechanical stability, and electrical and optical effects (Alen *et al.*, 2019). As mentioned previously, graphite is oxidized, forming graphite oxide which is contained with oxygen functional groups, including epoxy, carbonyl, carboxyl, and hydroxyl groups. This is then exfoliated to form GO. Experiments have been conducted on several successful GO-based gas separation membrane technologies, and it has been proposed that laminar structures and controlled structural flaws on GO layers may control gas separation performance and efficiency.

Graphene is considered as an ideal membrane platform due to it being only one atom thick. It was reported that a GO interlayer allows only water molecules to permeate, and tailoring surface groups can increase CO_2 permeability and selectivity, making it useful in CCUS. Wang *et al.* modified graphene oxide nanochannel or interlayer distance using borate as a cross-linker (Wang *et al.*, 2016). This manipulation increases CO_2 permeance to 650 GPU and selectivity over CH_4 (CO_2/CH_4) to 75. Borate crosslinking at 70°C changed nanochannel size from 0.44 to 0.35 nm, between the kinetic diameters of CO_2 and CH_4, which are 0.33 and 0.38 nm, respectively. Sieving is thus improved, resulting in slightly decreased permeance and increased selectivity. In another report, defects are introduced on the surface of a single layer of graphene by subjecting the layer to oxygen plasma treatment (He *et al.*, 2019a), allowing the graphene layer to be functionalized by polyethylenimine (PEI), which is CO_2-philic. This study reports CO_2 permeance of 6180 GPU and selectivity of 22.5 (CO_2/N_2). Another CO_2-philic, ethylenediamine (EDA), was used by the Yu group to improve the permeance and selectivity of GO for CO_2 (Zhou *et al.*, 2019). A membrane was developed by coating EDA functionalized

GO on the inner surface of poly (ether sulfone) (PES) hollow fiber. CO_2 permeance and CO_2/N_2 selectivity of 660 GPU and 572, respectively, were achieved in the wet condition at 75°C.

GO can be utilized as filler in MMMs due to its high aspect ratio, tunable surface functionality, and high mechanical and thermal properties (Huang *et al.*, 2018). The high aspect ratio provides longer and difficult paths for larger gas membranes, reducing permeability through the polymer matrix. The tunable surface functionality could simultaneously provide selective pathways for gas separation between GO nanolayers and improve polymer matrix compatibility through the presence of polar groups, such as hydroxyl, carbonyl, and epoxide. Huang *et al.* incorporated imidazolium-based ionic liquid (IL) into GO to synthesize Pebax/GO-IL MMMs to improve CO_2 solubility due to the IL's basic nature and superior CO_2-philic properties (Huang *et al.*, 2018).

GO can be made into an ultrathin membrane, exhibiting excellent CO_2/N_2 selectivity and permeability in several studies (Zhou *et al.*, 2019). Addition of an effective CO_2-philic carrier, such as a primary amine group, would drastically improve both CO_2 permeability and gas selectivity. Accordingly, Zhou *et al.* introduced ethylenediamine (EDA), a primary amine, to an ultrathin GO membrane through solution dispersion and modified vacuum-assisted coating (Zhou *et al.*, 2019). The EDA was successfully incorporated into GO sheets through crosslinking during synthesis. The EDA in the GO would act as a carrier agent for CO_2 through a reversible reaction from CO_2 into a bicarbonate (HCO_3) complex on the membrane. The complex would eventually migrate across the membrane and be converted back into CO_2. The primary amine would promote the HCO_3- reaction and induce rapid CO_2 diffusion into the membrane. The facilitated transport dramatically increased both CO_2 selectivity and permeability, improving GO carbon capture feasibility (Zhou *et al.*, 2019).

2.2.4. *Graphene-based mixed matrix membranes*

As discussed earlier, polymeric-based membranes are popular as they are highly permeable, inexpensive, and highly scalable. They suffer from low selectivity and low chemical and thermal stability. Conversely, inorganic-based membranes have molecular sieving properties that offer high selectivity. It can also be applied at relatively high temperatures and offers high chemical and mechanical stability, contributing to long life spans. High production costs and restricted scalability limit their use in CCUS, however. Utilization of inorganic materials such as graphene-based materials as filler in MMMs offers synergies resulting in high selectivity and permeability. One drawback to MMMs is non-ideal interfacial morphologies between filler and polymer matrix. Several studies report that functionalized GO improves interfacial interactions with organic polymers by way of addressing the problem. Organosilane-modified GO was developed to increase the chain mobility of the poly(ether-block-amide) (PEBAX) matrix (Zhang *et al.*, 2019). In this case, the silane group increases the degree of GO dispersion in the PEBAX matrix, also increasing CO_2 separation performance, where 0.9% of functionalized GO in the PEBAX matrix displays CO_2 permeance of 934.3 Barrer with 40.9 and 71.1 CO_2/CH_4 and CO_2/N_2 selectivity, respectively.

GO filler dispersibility in the matrix can also be increased by introducing the IL on the GO surface, with IL with high affinity to CO_2 preferred, as such also confering CO_2 permeance and selectivity. Huang *et al.* modified GO with an IL of 1-(3-aminopropyl)-3-methylimidazolium bromide, dispersed in PEBAX 1657 (Huang *et al.*, 2018). The IL also elevated CO_2 permeance to 900 GPU and CO_2/N_2 and CO_2/H_2 selectivity to 45 and 5.8, respectively. The GO filler was also modified by a sulfonate polymer brush to increase interface compatibility with a SPEEK polymer membrane (Xin *et al.*, 2019), increasing free volume cavity radius and fractional volume as well as interface compatibility, widening the CO_2 transport pathway. This allows increasing permeability to 1327 Barrer and CO_2-philic sites due to the presence of polymer brush CO_2/CH_4 selectivity by 179%.

Various approaches have been conducted relating to the development of graphene for CO_2 capture. Saleh proposed using nanomaterials and hybrid nanocomposites for CO_2 capture and utilization concerning environmental and energy sustainability (Saleh, 2022). Dong *et al.* developed GO nanosheet-based novel that facilitated transport membranes (Dong *et al.*, 2016). He *et al.* developed a porous GO/porous organic polymer hybrid nanosheet-functionalized MMM (He *et al.*, 2019b). Asghari *et al.* developed Polypyrrole-aided surface modification of GO nanosheets as fillers for poly(ether-b-amid) MMMs (Asghari *et al.*, 2021). Chowdhury and Balasubramanian developed holey graphene frameworks for highly selective post-combustion carbon capture (Chowdhury & Balasubramanian, 2016). Nazrul *et al.* developed arginine-containing chitosan-GO aerogels for highly efficient carbon capture and fixation (Hsan *et al.*, 2022). Thiago *et al.* proposed experimental and theoretical studies for CO_2 and H_2 adsorption on 3D nitrogen-doped porous graphene (dos Santos *et al.*, 2021). Giulia assessed the adsorption of CO_2 gas on graphene–polymer composites (Meconi *et al.*, 2019). Micari *et al.* reported on techno-economic assessments of post-combustion carbon capture using high-performance nanoporous single-layer graphene membranes (Micari *et al.*, 2021). Swati *et al.* reviewed CO_2 separation utilizing membrane technologies targeting capture from different emission sources and purification prior to conversion or utilization (Singh *et al.*, 2021). Moreover, Xinye *et al.* proposed the transformation of CO_2 to graphene (Liu *et al.*, 2020b).

3. Methodology

3.1. *Preparation of GO membrane*

Briefly, 3 g expanded graphite (D50 ~ 15 μm) and 18 g potassium permanganate ($KMnO_4$) were added to a mixture of 360 mL H_2SO_4 and 40 mL H_3PO_4 (360:40), and heated in a constant state stirring in an oil bath at a temperature of 50°C for 12 hours. The solution was cooled to room temperature, then mixed with 400 mL ice water and 2–6 mL hydrogen peroxide until the solution was brownish-yellow. The product was centrifuged for 10 minutes at 4000 rpm, washed three times each with a five-percent HCl solution and distilled water, and then centrifuged again at 4000 rpm for an additional 20–30 minutes. Graphite oxide is added to distilled water which is sonicated for five hours at a temperature below 60°C, after which the solution is centrifuged at 10,000 rpm for a further 30 minutes. Furthermore, the GO solution was subjected to the vacuum filtration machine for membrane fabrication. The final membranes were stored at room temperature prior to characterization.

3.2. *Membrane performance testing*

We evaluated membrane performance versus CO with a CO gas apparatus. The membrane is placed inside a chamber having two channels, one on either side. One side is connected to the CO tube that supplies CO gas. The performance of the membrane is measured by detecting CO resistivity after exposure.

4. Results and Discussion

4.1. *Prospects, utilization, and challenges of membrane technology in CCUS*

In both ASEAN and East Asia, membrane technologies confront such challenges as high capital investment and low return on investment (ROI), especially in carbon storage (Lau, 2022). Different ASEAN member states would implement CCUS differently according to their respective plans for reaching carbon neutrality.

In East Asian nations such as China, CCUS demonstration projects have been cost-intensive, due to the high investment cost and low ROI, while cheap fossil fuels and underdeveloped CO_2 utilization technology make CCUS economically unfeasible (Chen *et al.*, 2022b). Applicable CCUS financial mechanisms are needed in addition to technological breakthroughs to drive more investment.

Studies of ASEAN CCUS directions by Lau *et al.* suggest that several measures should be taken to decarbonize ASEAN: increasing sustainable renewable energy sources in power generation, transitioning from coal to gas generation, electrification of road transport, hydrogen for marine transport, biofuel for aviation, blue hydrogen for industry, and CCUS corridors (Lau *et al.*, 2022). ASEAN should utilize CCUS to keep harnessing fossil fuels sustainably while simultaneously implementing more renewable energy, which can be done through post-combustion CCUS that uses fossil fuels which contribute the largest percentage of current ASEAN TPES. Saline aquifers contained in major sedimentary basins in ASEAN coastal regions have approximately 98% CO_2 storage capacity, which can be utilized for ASEAN CO_2 storage needs. CCUS corridors may also be formed across ASEAN to funnel captured CO_2 to common sinks or permanent storage, hopefully reducing CCUS costs as well as improving economic incentives for policy makers and public perception alike (Lau *et al.*, 2022). New technologies offering potential solutions to CCUS problems will still be needed.

4.2. *Analysis of membrane technology utilization*

The technology readiness level (TRL) of membrane technology has increased over time, indicating substantial progress toward its CCUS feasibility. The low unit cost and modular setup are significant advantages, given the economic feasibility concerns. Breakthroughs such as the graphene-based material incorporation into polymeric membranes would improve performance while reducing complexity through simplified synthesis. More pilot tests and post-combustion carbon capture tests are necessary to drive membrane technology adoption in CCUS.

4.3. *Challenges to Membrane Utilization in CCUS*

Srivastav *et al.* conducted a CCUS study in Singapore, one of ASEAN's most developed member states (Srivastav *et al.*, 2021). They compiled CO_2 emission profile data from the country's energy and chemical sectors, and future implementations for several available CCUS technologies. The assessment covered capture by chemical absorption including benchmark proprietary MEA amine process (TRL 9), advanced amines or improved conventional solvents (TRL 9), chilled ammonia (TRL 6–8), potassium carbonate (TRL 4–5), solid physical adsorption including temperature swing adsorption (TRL 7), membrane separation including polymeric membrane (TRL 6) with the advantage of fast development due to the modular setup, chemical looping including calcium carbonate looping in cement production (TRL 6), and cryogenic distillation (TRL 4) among others.

Even though membrane technology has low TRL numbers, cost reduction potential and fast development due to extensive research make it a promising CCUS technology (Srivastav *et al.*, 2021). As of 2022, its TRL score varied between 6 and 9 (Chen *et al.*, 2022b), showing significant progress toward active use in fossil fuel plants (Srivastav *et al.*, 2021).

A major challenge to membrane technology in CCUS is competition with other technologies, as it requires more research to be made viable in post-combustion carbon capture in working fossil fuel plants. More efforts are also needed for CUS, as these applications are still limited, and membrane technology remains a challenge.

4.4. *Experimental Analysis of GBM Structure*

Figure 1 shows a GO membrane, with a 2 cm average diameter and a shiny surface. A cross-section SEM was used to confirm GO formation. The membrane was assembled from layered graphene having a 6 μm of

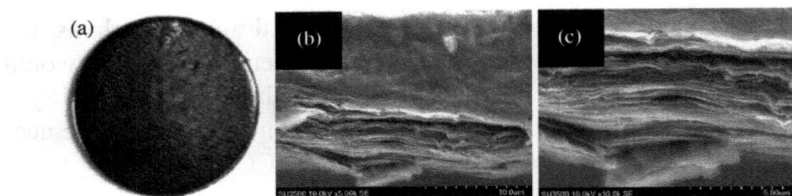

Figure 1.   Graphene-based membrane (a), cross-section SEM image of GO (b), and magnified laminated GO layer (c).

total thickness. The interlayer distance is observed to be large enough to allow molecules such as CO or CO_2 to move and reach deeper graphene parts, increasing storage capacity.

4.5. *Experimental Analysis of Membrane Performance Versus Gaseous Compounds*

As mentioned earlier, CO_2 molecules and other pollutant gases created by industry and transportation are then supposed to be captured by CCUS. Gas sensor technology can be utilized to measure amounts of released molecules and other dangerous compounds such as carbon monoxide (CO) created from incomplete reactions. GO-based membranes have been evaluated to detect CO using chemiresistive sensing. Active functional groups such as carboxyl, hydroxyl, and carbonyl on GO surfaces easily interact with carbon-based molecules like CO. Figure 2 shows the dynamic response of a GO membrane to CO molecules at room temperature. The membrane possesses resistance of 0.44 MΩ in a normal atmosphere, increasing as the membrane is exposed to CO at concentrations of 25 ppm, implying that the electronic properties of the membrane change in response to the presence of CO. At room temperature, CO adsorbs on the surface, interacting with the groups that existed on the membrane surface as mentioned above (Ha *et al.*, 2018; Mohammed *et al.*, 2022; Panda *et al.*, 2016). The mechanism is strengthened by the fact that the sensing response increases as the amount of CO rises to 50 ppm, indicating more molecules disrupting the charge distribution. These results are anticipated to be explored for CO_2 capture by using graphene-based membranes.

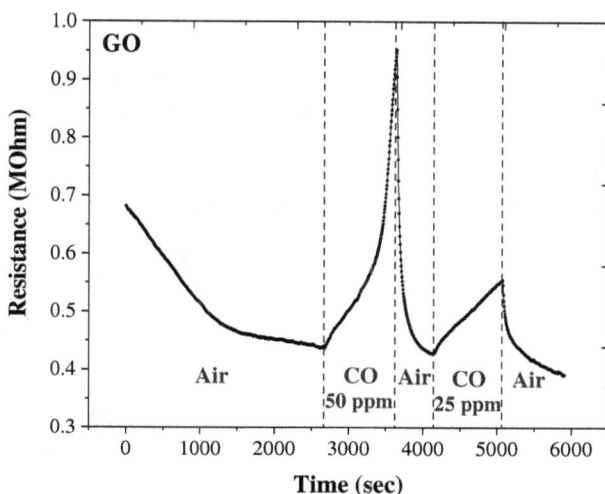

Figure 2. Gas sensing performance of graphene-based membranes.

5. Conclusion and Policy Implications

In summary, Indonesia is now conducting 15 CCS/CCUS projects that are in the study phase. The role of international collaboration is crucial as Indonesia receiving assistance from developed countries like Japan will help further its CCUS initiatives. One of the CCUS methods involves the use of membrane technology. In this study, graphene-based membranes have been successfully prepared by vacuum filtration, having a layered structure with a large enough interlayer distance to facilitate gas intercalation or provide many active sites for gas moiety. The CO sensing performance of the membrane has also been evaluated, showing a significant response to 25 ppm of CO at room temperature. The results show that GBM has multiple functions. Based on our study, the exceptional properties of GBM are believed to work very well in gas or CO capture, contributing to the reduction of the global carbon footprint. In relation to policies in Indonesia, the results of this activity strongly support the national commitment to realize the Paris Agreement to the United Nations Framework Convention on Climate Change to achieve zero greenhouse gas emissions and climate resilience by 2050.

References

Al-Mamoori, A., Krishnamurthy, A., Rownaghi, A. A., and Rezaei, F. (2017). Carbon capture and utilization update. *Energy Technology*, *5*(6), 834–849. https://doi.org/10.1002/ente.201600747.

Asghari, M., Saadatmandi, S., and Parnian, M. J. (2021). Polypyrrole-aided surface decoration of graphene oxide nanosheets as fillers for poly(ether-b-amid) mixed matrix membranes to enhance CO_2 capture. *International Journal of Energy Research*, *45*(7), 10843–10857. https://doi.org/10.1002/er.6567.

Babu, D. J., He, G., Hao, J., Vahdat, M. T., Schouwink, P. A., Mensi, M., and Agrawal, K. V. (2019). Restricting lattice flexibility in polycrystalline metal–organic framework membranes for carbon capture. *Advanced Materials*, *31*(28), 1900855. https://doi.org/10.1002/adma.201900855.

Chen, G., Wang, T., Zhang, G., Liu, G., and Jin, W. (2022a). Membrane materials targeting carbon capture and utilization. *Advanced Membranes*, *2*, 100025. https://doi.org/10.1016/j.advmem.2022.100025.

Chen, S., Liu, J., Zhang, Q., Teng, F., and McLellan, B. C. (2022b). A critical review on deployment planning and risk analysis of carbon capture, utilization, and storage (CCUS) toward carbon neutrality. *Renewable & Sustainable Energy Reviews*, *167*, 112537. https://doi.org/10.1016/j.rser.2022.112537.

Chernikova, V., Shekhah, O., Belmabkhout, Y., and Eddaoudi, M. (2020). Nanoporous fluorinated metal–organic framework-based membranes for CO_2 capture. *ACS Applied Nano Materials*, *3*(7), 6432–6439. https://doi:10.1021/acsanm.0c00909.

Chowdhury, S. and Balasubramanian, R. (2016). Holey graphene frameworks for highly selective post-combustion carbon capture. *Scientific Reports*, *6*(1), 21537. https://doi:10.1038/srep21537.

Dong, G., Zhang, Y., Hou, J., Shen, J., and Chen, V. (2016). Graphene oxide nanosheets based novel facilitated transport membranes for efficient CO_2 capture. *Industrial & Engineering Chemistry Research*, *55*(18), 5403–5414. https://doi:10.1021/acs.iecr.6b01005.

Dos Santos, T. C., Mancera, R. C., Rocha, M. V. J., da Silva, A. F. M., Furtado, I. O., Barreto, J., Stavale, F., Archanjo, B. S., Carneiro, J. W. d. M., Costa, L. T., and Ronconi, C. M. (2021). CO_2 and H_2 adsorption on 3D nitrogen-doped porous graphene: Experimental and theoretical studies. *Journal of CO_2 Utilization*, *48*, 101517. https://doi.org/10.1016/j.jcou.2021.101517.

Elsaidi, S. K., Ostwal, M., Zhu, L., Sekizkardes, A., Mohamed, M. H., Gipple, M., McCutcheon, J. R., and Hopkinson, D. (2021). 3D printed MOF-based mixed matrix thin-film composite membranes. *RSC Advances*, *11*(41), 25658–25663. https:// doi:10.1039/D1RA03124D.

Favre, E. (2022). Membrane separation processes and post-combustion carbon capture: State of the art and prospects. *Membranes (Basel)*, *12*(9). https://doi:10.3390/membranes12090884.

Gizer, S. G., Polat, O., Ram, M. K., and Sahiner, N. (2022). Recent developments in CO_2 capture, utilization, related materials, and challenges. *International Journal of Energy Research*, *46*(12), 16241–16263. https://doi.org/10.1002/er.8347.

Hasan, M. F., First, E. L., Boukouvala, F., and Floudas, C. A. (2015). A multi-scale framework for CO_2 capture, utilization, and sequestration: CCUS and CCU. *Computers & Chemical Engineering*, *81*, 2–21. https://doi.org/10.1016/j.compchemeng.2015.04.034.

He, G., Huang, S., Villalobos, L. F., Zhao, J., Mensi, M., Oveisi, E., Rezaei, M., and Agrawal, K. V. (2019). High-permeance polymer-functionalized single-layer graphene membranes that surpass the post combustion carbon capture target. *Energy & Environmental Science*, *12*(11), 3305–3312. https://doi:10.1039/C9EE01238A.

He, R., Cong, S., Wang, J., Liu, J., and Zhang, Y. (2019). Porous graphene oxide/porous organic polymer hybrid nanosheets functionalized mixed matrix membrane for efficient CO_2 capture. *ACS Applied Materials & Interfaces*, *11*(4), 4338–4344. https://doi:10.1021/acsami.8b17599.

He, X. (2018a). A review of material development in the field of carbon capture and the application of membrane-based processes in power plants and energy-intensive industries. *Energy, Sustainability & Society*, *8*(1), 1–14. https://doi.org/10.1186/s13705-018-0177-9.

He, X. (2018b). A review of material development in the field of carbon capture and the application of membrane-based processes in power plants and energy-intensive industries. *Energy, Sustainability and Society*, *8*(1), 34. https://doi:10.1186/s13705-018-0177-9.

Hou, R., Fong, C., Freeman, B. D., Hill, M. R., and Xie, Z. (2022). Current status and advances in membrane technology for carbon capture. *Separation and Purification Technology*, *300*, 121863. https://doi.org/10.1016/j.seppur.2022.121863.

Hsan, N., Dutta, P. K., Kumar, S., and Koh, J. (2022). Arginine containing chitosan-graphene oxide aerogels for highly efficient carbon capture and fixation. *Journal of CO_2 Utilization*, *59*, 101958. https://doi.org/10.1016/j.jcou.2022.101958.

Huang, G., Isfahani, A. P., Muchtar, A., Sakurai, K., Shrestha, B. B., Qin, D., Yamaguchi, D., Sivaniah, E., and Ghalei, B. (2018). Pebax/ionic liquid modified graphene oxide mixed matrix membranes for enhanced CO_2 capture. *Journal of Membrane Science*, *565*, 370–379. https://doi.org/10.1016/j.memsci.2018.08.026.

Stavins, R., Zou, J., Brewer, T., Conte Grand, M., den Elzen, M., Finus, M., Gupta, J., Höhne, N., Lee, M.-K., Michaelowa, A., Paterson, M., Ramakrishna, K., Wen, G., Wiener, J., and Winkler, H. (2014). *International Cooperation: Agreements and Instruments. In: Climate Change 2014: Mitigation of Climate Change. Contribution of Working Group III to the Fifth Assessment Report of the Intergovernmental Panel on Climate Change* [Edenhofer, O., R. Pichs-Madruga, Y. Sokona, E. Farahani, S. Kadner, K. Seyboth, A. Adler, I. Baum, S. Brunner, P. Eickemeier, B. Kriemann, J. Savolainen, S. Schlömer, C. von Stechow, T. Zwickel and J.C. Minx (eds.)]. Cambridge University Press, Cambridge, United Kingdom and New York, NY, USA.

Kárászová, M., Zach, B., Petrusová, Z., Červenka, V., Bobák, M., Šyc, M., and Izák, P. (2020). Post-combustion carbon capture by membrane separation, Review. *Separation & Purification Technology, 238*, 116448. https://doi.org/10.1016/j.seppur.2019.116448.

Kawakami, M., Iwanaga, H., Hara, Y., Iwamoto, M., and Kagawa, S. (1982). Gas permeabilities of cellulose nitrate/poly(ethylene glycol) blend membranes. *Journal of Applied Polymer Science, 27*(7), 2387–2393. https://doi.org/10.1002/app.1982.070270708.

Kentish, S. E. (2019). 110th anniversary: Process developments in carbon dioxide capture using membrane technology. *Industrial & Engineering Chemistry Research, 58*(28), 12868–12875. https://doi:10.1021/acs.iecr.9b02013.

Lau, H. C. (2022). Decarbonization roadmaps for ASEAN and their implications. *Energy Reports, 8*, 6000–6022. https://doi.org/10.1016/j.egyr.2022.04.047.

Lau, H. C., Zhang, K., Bokka, H. K., and Ramakrishna, S. (2022). A review of the status of fossil and renewable energies in Southeast Asia and its implications on the decarbonization of ASEAN. *Energies, 15*(6), 2152. https://doi.org/10.3390/en15062152.

Liang, C. Z., Yong, W. F., and Chung, T.-S. (2017). High-performance composite hollow fiber membrane for flue gas and air separations. *Journal of Membrane Science, 541*, 367–377. https://doi.org/10.1016/j.memsci.2017.07.014.

Liu, M., Nothling, M. D., Webley, P. A., Jin, J., Fu, Q., and Qiao, G. G. (2020a). High-throughput CO_2 capture using PIM-1@MOF based thin film composite membranes. *Chemical Engineering Journal, 396*, 125328. https://doi.org/10.1016/j.cej.2020.125328.

Liu, X., Wang, X., Licht, G., and Licht, S. (2020b). Transformation of the greenhouse gas carbon dioxide to graphene. *Journal of CO_2 Utilization, 36*, 288–294. https://doi.org/10.1016/j.jcou.2019.11.019.

Meconi, G. M., Tomovska, R., and Zangi, R. (2019). Adsorption of CO_2 gas on graphene–polymer composites. *Journal of CO_2 Utilization, 32*, 92–105. https://doi.org/10.1016/j.jcou.2019.03.005.

Micari, M., Dakhchoune, M., and Agrawal, K. V. (2021). Techno-economic assessment of postcombustion carbon capture using high-performance nanoporous single-layer graphene membranes. *Journal of Membrane Science, 624,* 119103. https://doi.org/10.1016/j.memsci.2021.119103.

Niwa, M., Kawakami, H., Nagaoka, S., Kanamori, T., and Shinbo, T. (2000). Fabrication of an asymmetric polyimide hollow fiber with a defect-free surface skin layer. *Journal of Membrane Science, 171*(2), 253–261. https://doi.org/10.1016/S0376-7388(00)00306-9.

Park, H. B. (2014). Graphene-based membranes — A new opportunity for CO_2 separation. *Carbon Management, 5*(3), 251–253. https://doi:10.1080/17583004.2014.923237.

Powell, C. E., and Qiao, G. G. (2006). Polymeric CO_2/N_2 gas separation membranes for the capture of carbon dioxide from power plant flue gases. *Journal of Membrane Science, 279*(1), 1–49. https://doi.org/10.1016/j.memsci.2005.12.062.

Saleh, T. A. (2022). Nanomaterials and hybrid nanocomposites for CO_2 capture and utilization: Environmental and energy sustainability. *RSC Advances, 12*(37), 23869–23888. https://doi:10.1039/D2RA03242B.

Shah, C., Raut, S., Kacha, H., Patel, H., and Shah, M. (2021). Carbon capture using membrane-based materials and its utilization pathways. *Chemical Papers, 75*(9), 4413–4429. doi:10.1007/s11696-021-01674-z.

Shi, L., Lai, L. S., Tay, W. H., Yeap, S. P., and Yeong, Y. F. (2022). Membrane fabrication for carbon dioxide separation: A critical review. *ChemBioEng Reviews.* https://doi.org/10.1002/cben.202200035.

Singh, S., Varghese, A. M., Reinalda, D., and Karanikolos, G. N. (2021). Graphene-based membranes for carbon dioxide separation. *Journal of CO_2 Utilization, 49,* 101544. https://doi.org/10.1016/j.jcou.2021.101544.

Soujanya, Y. (2020). Chapter 10 — CO_2 adsorption by functionalized sorbents. In M. R. Rahimpour, M. Farsi, & M. A. Makarem (Eds.), *Advances in Carbon Capture* (pp. 229–240). Woodhead Publishing, Duxford, United Kingdom.

Srivastav, P., Schenkel, M., Mir, G. R., Berg, T., and Staats, M. (2021). Carbon capture, utilisation and storage (CCUS): Decarbonisation pathways for Singapore's energy and chemicals sectors. *Navigant Netherlands B.V.* Retrieved from https://file.go.gov.sg/carbon-capture-utilisation-and-storage-decarbonisation-pathway-for-singapore-energy-and-chemical-sectors-pdf.pdf.

Sule, M. R. and Widiyantoro, S. (2018). Gundih carbon capture and storage pilot project: Current status of the first CCS project in South and Southeast Asian Regions. *Presentation at Blue Carbon Summit at The National Library of Indonesia,* Jakarta, 17–18 July 2018, pages 1–22. Available at https://www.cifor-icraf.org/knowledge/slide/12523/.

The Ministry of Energy and Mineral Resources (MEMR). (2022). Progress of CCS/CCUS implementation in Indonesia by Dr. Muhammad Wafid Agung (The Ministry of Energy and Mineral Resources of Republic of Indonesia) presented at The Third Asia CCUS Network Forum 27 September 2023, at Hiroshima and Hybrid. Page 1–6. Available at https://acnf.jp/program/file/presentation/7_Panelist_MEMR-IDN.pdf.

UNECE. (2021). *Technology Brief: Carbon Capture, Use and Storage (CCUS).* New York: United Nations Economic Commission for Europe.

Venna, S. R., and Carreon, M. A. (2015). Metal organic framework membranes for carbon dioxide separation. *Chemical Engineering Science, 124,* 3–19. https://doi.org/10.1016/j.ces.2014.10.007.

Wang, S., Wu, Y., Zhang, N., He, G., Xin, Q., Wu, X., Wu, H., Cao, X., Guiver, M. D., and Jiang, Z. (2016). A highly permeable graphene oxide membrane with fast and selective transport nanochannels for efficient carbon capture. *Energy & Environmental Science, 9*(10), 3107–3112. https://doi:10.1039/C6EE01984F.

Wu, W., Li, Z., Chen, Y., and Li, W. (2019). Polydopamine-modified metal–organic framework membrane with enhanced selectivity for carbon capture. *Environmental Science & Technology, 53*(7), 3764–3772. https://doi:10.1021/acs.est.9b00408.

Xin, Q., Ma, F., Zhang, L., Wang, S., Li, Y., Ye, H., … Cao, X. (2019). Interface engineering of mixed matrix membrane via CO_2-philic polymer brush functionalized graphene oxide nanosheets for efficient gas separation. *Journal of Membrane Science, 586,* 23–33. https://doi.org/10.1016/j.memsci.2019.05.050.

Yu, K. M. K., Curcic, I., Gabriel, J., and Tsang, S. C. E. (2008). Recent advances in CO_2 capture and utilization. *ChemSusChem, 1*(11), 893–899. https://doi.org/10.1002/cssc.200800169.

Zhang, J., Xin, Q., Li, X., Yun, M., Xu, R., Wang, S., Li, Y., Lin, L., Ding, X., Ye, H., and Zhang, Y. (2019). Mixed matrix membranes comprising aminosilane-functionalized graphene oxide for enhanced CO_2 separation. *Journal of Membrane Science, 570–571,* 343–354. https://doi.org/10.1016/j.memsci.2018.10.075.

Zhou, F., Tien, H. N., Dong, Q., Xu, W. L., Li, H., Li, S., and Yu, M. (2019). Ultrathin, ethylenediamine-functionalized graphene oxide membranes on hollow fibers for CO_2 capture. *Journal of Membrane Science, 573,* 184–191. https://doi.org/10.1016/j.memsci.2018.11.080.

Chapter 7

Carbon Storage in Harvested Products along the Wood Supply Chain through Industry 4.0 Technologies: Challenges, Opportunities, and Policy Implications

Nophea Sasaki

Natural Resources Management, Asian Institute of Technology, Bangkok, Thailand

nopheask@gmail.com

Abstract

The global wood industry annually harvests and consumes approximately 2.2 billion m³, presenting a significant opportunity to increase carbon storage within harvested wood products (HWPs) through strategic implementation of advanced technologies. While the advent of the Industry 4.0 revolution has introduced a suite of technologies with the potential to enhance carbon storage in HWPs, research has been limited on this crucial subject. This chapter provides a comprehensive review of past efficiencies in harvesting, processing, and manufacturing along the supply chain within the Southeast Asian wood industry and proposes well-informed policy recommendations based on the research to facilitate the early adoption of cutting-edge technologies, with the aim of optimizing forest utilization and management while increasing carbon storage in HWPs. This chapter advocates for sustainable practices that align with global climate goals and promote a greener future by addressing the critical knowledge gap regarding the utilization of Industry 4.0 technologies to enhance carbon storage in HWPs.

Keywords: Timber harvesting, half-life time, carbon storage, carbon incentives, sustainability, wood manufacturing, wood supply chain, green financing.

1. Introduction

Global efforts concerning the Sustainable Development Goals (SDGs), established by the United Nations in 2015, have focused on achieving sustainability by 2030 (United Nations Economic and Social Commission for Asia and the Pacific, 2022). While progress has been made on certain SDGs, challenges remain in areas such as climate action (SDG13) and responsible and sustainable consumption (SDG12), particularly in the post-COVID-19 era. The adverse impacts of climate change and ongoing depletion of forest resources have immediate and long-term consequences for economic and ecosystem stability, necessitating urgent measures to mitigate climate change, conserve forests, and promote responsible consumption practices. In recent years, carbon capture, utilization, and storage (CCUS) has emerged as a key strategy to address climate change.

With their potential for carbon storage, forests play a crucial role in CCUS, depending on sustainable management practices. Unfortunately, mismanagement and inefficient utilization of forest resources have resulted in huge deforestation and forest degradation, causing up to 25 percent of global carbon emissions. Therefore, sustainable utilization of forest resources is vital for addressing the climate change. The advent of the Industry 4.0 revolution in 2011 has opened up new possibilities for enhancing CCUS through improved utilization of forest resources. The nine technologies associated with Industry 4.0 offer opportunities to optimize carbon storage along the wood supply chain while promoting sustainable practices and efficient resource utilization (Ramos-Maldonado and Aguilera-Carrasco, 2021).

The wood industry holds global significance as it meets increasing demand for timber, creates employment opportunities, and has the potential to contribute to biodiversity conservation through effective management practices. Unfortunately, unsustainable logging practices, inadequate processing, and manufacturing technologies have resulted in significant wood waste, hindering the full potential of CCUS. Studies by Sasaki *et al.* (2012) and (2016) have demonstrated that implementing technologies that enhance efficiency along the wood industry's supply chain can substantially reduce waste, increasing carbon storage potential. However, research on the application of the nine technologies of the Industry 4.0 within the wood industry has been limited, particularly regarding CCUS, making it difficult to suggest policy interventions regarding the use of these technologies for the CCUS.

The aims of this study are to (1) assess application of Industry 4.0 technologies in increasing carbon storage within harvested wood products (HWPs) by improving efficiency, reducing waste, and promoting long-term viability of manufactured wood products as carbon sinks; (2) provide policy recommendations for early adoption of Industry 4.0 technologies in Southeast Asia's wood industry, focusing on CCUS; (3) explore carbon pricing policies that give incentives to adopting Industry 4.0 technologies.

By examining the potential of Industry 4.0 technologies to maximize carbon storage along the wood supply chain, this research can contribute to CCUS research and global climate change mitigation efforts. Findings will offer basis to policymakers and industry stakeholders in Southeast Asia, facilitating adopting sustainable practices that promote environmental conservation, economic prosperity, and effective utilization of carbon storage capacities in the wood industry.

2. Analysis of the Industry 4.0 Technologies and Wood Industry Applications

2.1. *Forest management and supply chain of HWPs*

Forest management encompasses strategic planning and implementation of practices to utilize forest resources, aiming to achieve environmental, social, and economic objectives (FAO, 2020b). The classification of wood products, as illustrated in Figure 1, indicates that a portion of harvested wood is transformed into such "semi-finished" products as sawnwood, panels, paper, and paperboard, which are subsequently utilized as primary wood products. Another portion is used as fuel. The residence times of primary wood products can vary, with some being short-lived, i.e., less

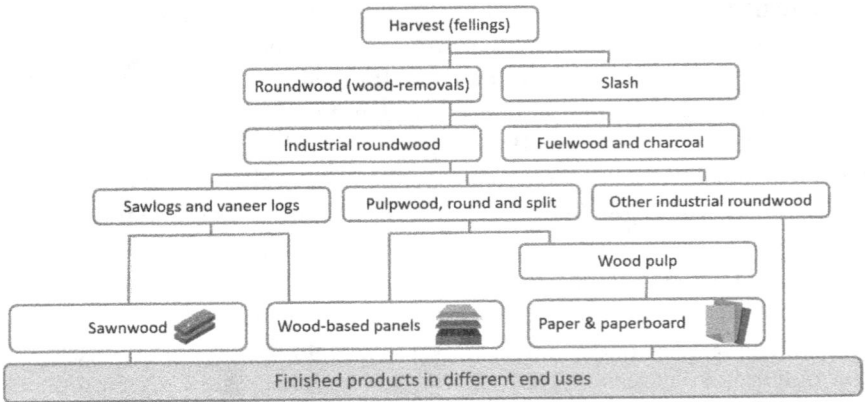

Figure 1.   Classification of wood products based on FAO forest products' definitions.
Source: IPCC (2014).

than 25 years, furniture and panels, ranging from 25 to 43 years, and construction wood, up to 50 years (Profft *et al.*, 2009). These differences in residence time influence the carbon stored in HWPs.

Müller *et al.* (2019a) identified five main stages in the wood supply chain: harvest planning, harvest organization and control, harvest operation, timber transport and logistics, and timber sales. Forest management operations, including harvest and transport, as well as supplying such forest products as paper, fiber boards, biomass, energy, and ethanol, constitute the primary entities within the forest supply chain (Meyer *et al.*, 2019). Figure 2 illustrates the key components of a forest product supply chain.

After harvest, roundwood is transported to sawmills, where logs are further processed to create value-added products (Borz and Păun, 2020). As the wood supply chain involves several operating units, including forest enterprise proprietors, transporters, pulp and sawmills, markets, and customers, encompassing sustainability can be difficult without efficient collaboration (Dalalah *et al.*, 2022).

In recent years, preservation and enhancement of forest carbon storage capacity has become an integral aspect of forest management. Ontl *et al.* (2020) proposed various forest management strategies to improve carbon storage and enhance ecosystem functions, including utilizing forwarders instead of skidders, locating landing sites near roads rather than

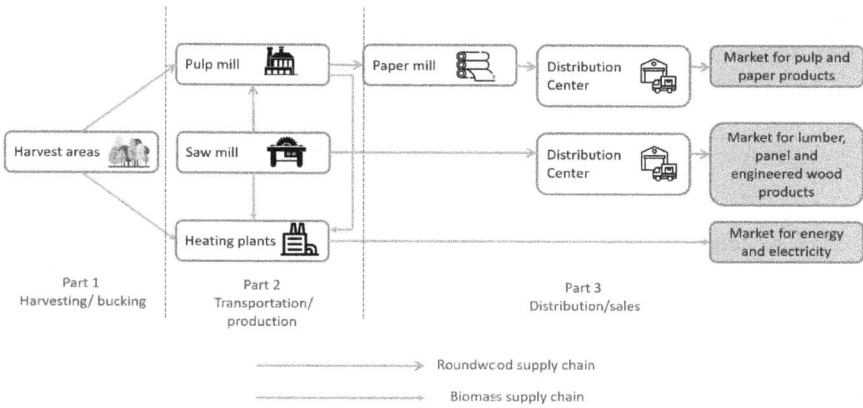

Figure 2. Key forest product supply chain components.
Source: Kong (2014).

within forests, preserving highly carbon-dense forests, implementing single-tree selection, crop tree release, and thinning. Similarly, improved forest management (IFM) practices involve longer rotation periods, thinning, promoting uneven-aged management, minimizing logging damage, selecting species strategically to maintain diversity, retaining coarse wood debris in stands, and reducing soil damage (Kaarakka *et al.*, 2021). A study found that effective forest management, coupled with efficient transfer of forest carbon to the wood product carbon pool, can contribute to long-term carbon storage (Liu and Han, 2009).

Dalalah *et al.* (2022) highlighted two aspects that would help make the wood supply chain more sustainable: recycling, and progressive designing or manufacturing to maximize utilization and minimalize waste. The authors emphasized that such Industry 4.0 technologies as smart factories could also help. Another study explored the economic, environmental, and social aspects of sustainability in wood supply chains, reporting that these focused on economic and environmental dimensions, and comparatively less on social aspects (Santos *et al.*, 2019). Studies focusing on wood supply chain economics mostly considered net present value (NPV) to maximize profit, whereas those focusing on environmental aspects analyzed greenhouse gas emissions (GHGs) and global warming potential (GWP) using Life Cycle Assessment (LCA) methods. Their study also found that while improving supply chains involved more strategic decisions, i.e., long-term planning, tactical decisions, related to inventory control, and transportation and production planning were also gaining impetus. Similarly, He and Turner (2021) reported that Industry 4.0 technologies used in wood supply chains were also more focused on lowering costs and increasing profits than social aspects.

2.2. *Current wood demand and supply*

A significant portion of global roundwood demand, including industrial roundwood and fuelwood, is sourced from plantation forests (Mishra *et al.*, 2021). Plantations, despite covering a smaller land area, are able to meet a substantial portion of roundwood demand due to intensive management practices (Nepal *et al.*, 2019). While roundwood can also be obtained from natural forests, managed secondary forests, or planted forests (Mishra *et al.*, 2021), ever-increasing demand has led to competition between wood production and other land use systems, including agriculture. This exacerbates land scarcity and drives deforestation (Mishra *et al.*, 2021). The construction sector's demand for timber is expected to continue rising due to its status as a bio-based, low-carbon material, making it more sustainable in the face of climate change than fossil fuel-based alternatives (Nepal *et al.*, 2021). Table 1 lists the countries producing the

Table 1.   Highest industrial roundwood producing countries.

No.	Country	Industrial Roundwood Production (million m^3)
1	Brazil	131.9
2	United States of America	101.9
3	China	64.2
4	India	43.1
5	Chile	38.4
6	New Zealand	27.5
7	Australia	19.2
8	South Africa	15.9
9	Thailand	14.6
10	Indonesia	12.5

Source: Jürgensen *et al.* (2014).

most industrial roundwood from plantation forests, collectively accounting for 83 percent of global production from such sources (Jürgensen
et al., 2014).

As climate change impacts are projected to drive more global timber
harvests and lower timber prices (Tian *et al.*, 2016), the growing demand
for roundwood, estimated to reach 6 billion m³ by 2050, is a significant
driver for the establishment of more plantation forests (Barua *et al.*,
2014). Expansion of such plantations is thus expected, particularly in
tropical and subtropical regions of Southeast Asia, Africa, and South
America (Mcewan *et al.*, 2020). Mishra *et al.* (2021) suggests that plantation areas could increase by 171 million ha by 2100 to meet rising
demand. According to the Food and Agriculture Organization of the
United Nations (FAO), plantation forests currently cover only 3 percent of
the global forest area (131 million ha) and are primarily managed for timber and other resource extraction (FAO, 2020a).

2.3. *Conventional technologies and their effects on efficiency in wood harvesting, processing, and manufacturing*

Until recently, motor-manual tree felling was one of the most used methods in conventional harvesting, in which manual operators equipped with
chainsaws are employed to fell trees. This approach is typically preferred
for larger trees and in areas where challenging terrain hinders use of
mechanized equipment (Ghaffariyan, 2021). Chainsaws are also prevalent
in regions with low labor costs, as they are more affordable than mechanized alternatives (Manavakun, 2014). However, motor-manual tree felling poses significant risks, including injuries and fatalities (Michael and
Gorucu, 2021). The force exerted by a falling tree can lead to uncontrolled
movement, causing damage to surrounding trees, soil, and other aspects
of the environment (Jourgholami *et al.*, 2013; Sasaki *et al.*, 2016). Other
study also indicates that inadequate maintenance of chainsaws can result
in decreased equipment performance, leading to increased fuel consumption (Kaakkurivaara and Stampfer, 2018).

A study by Kaakkurivaara and Stampfer (2018) demonstrated that
replacing manual timber extraction with skidder vehicles increased productivity from 2.13 m³ha⁻¹ to 6.25 m³ha⁻¹, while being 59 percent
more expensive. Methods such as tractors and portable winches have
also shown greater effectiveness than manual extraction. Therefore,

implementing advanced harvesting technologies can help reduce timber extraction costs and improve forest management practices, allowing revenue to cover expenses and making harvesting more cost-effective (Bont *et al.*, 2022).

2.4. *Effects of technologies on carbon storage in HWPs*

According to the IPCC (2006), all wood constituents leaving harvest sites are treated as HWPs, encompassing a wide range of products from furniture to fuel. HWPs can act as either sources or sinks of carbon, with carbon being added to the HWP pool during the manufacturing of new forest products and released when older products reach end-of-life and decay in landfills or are burned (IPCC, 2006; Johnston and Radeloff, 2019). The HWP carbon stock is influenced by durability, i.e., longevity, and recycling rate (Brunet-Navarro *et al.*, 2018). Carbon stored in HWPs derived from sustainably managed forests also contributes to SDG 8, promotion of sustainable economic growth, SDG 12, responsible consumption and production, SDG 13, combating climate change and its impacts, and SDG 15, protection, restoration, and sustainable use of terrestrial ecosystems (FAO, 2022). Expanding carbon storage outside forests through HWPs thus plays a significant role in climate change mitigation efforts and offers cost-efficient potential for negative emissions, aligning with CCUS goals (Iordan *et al.*, 2018).

As a bio-based material, HWPs are increasingly preferred over fossil fuel-based alternatives, particularly for their aforementioned potential in climate change mitigation and CCUS (Parobek *et al.*, 2019). The amount of carbon stored in HWPs depends on the product, as some have longer lifespans than others. For instance, sawnwood or panels can last from 10 to 100 years, while paper has a lifespan of less than 5 years, and fuelwood may be burned in the same year it is harvested (IPCC, 2006). Even if the harvested volume remains the same, HWP products with longer lifespans, such as sawnwood or panels, accumulate greater amounts of carbon (Parobek *et al.*, 2019). As mentioned above, maximizing carbon storage in HWPs thus aligns with CCUS objectives CCUS by effectively capturing and utilizing carbon in long-lived wood products.

One approach to increase carbon storage in HWPs is to enhance the growth rate of tree plantations, resulting in greater carbon accumulation in standing trees and wood products (Nepal *et al.*, 2012). Management intensity can be increased through such means as fertilization, pest control, and site preparation (Stantutf *et al.*, 2003). Peña-Claros *et al.* (2008) found that increasing logging intensity and implementing additional silvicultural practices, including removing lianas, eliminating competition through girdling, and reduced impact logging techniques (RIL), enhanced tree growth. Ligot *et al.* (2019) found that other logging and silvicultural treatments including thinning stimulated tree growth rates for up to 15 years in *Triplochiton scleroxylon* stands, increasing HWP carbon capture potential. Another strategy to increase carbon storage in HWPs is by extending the cutting rotation age of tree stands (Nepal *et al.*, 2012). Nepal *et al.* (2012) conducted a study on *Pinus taeda* stands, finding that increasing the cutting rotation by 5 years resulted in a 4 percent increase in HWP carbon storage. Extending the rotation by 65 years could potentially increase carbon storage by up to 26 percent.

Promoting HWPs with longer lifespans, such as the aforementioned sawnwood and panels, and encouraging recycling, can significantly enhance carbon storage (FAO, 2022; Brunet-Navarro *et al.*, 2017; Werner *et al.*, 2010). Shifting production toward longer-lived products and increasing the half-life of paper and paperboard products or extending the lifespan of sawnwood products through recycling contribute to higher carbon stocks in HWPs (Brunet-Navarro *et al.*, 2017; Paluš *et al.*, 2020). These strategies align with CCUS goals by maximizing carbon storage potential and promoting sustainable utilization of wood resources. Implementing these practices can support the transition to a low-carbon economy and contribute to global efforts in mitigating climate change through CCUS.

2.5. *Industry 4.0: Revolutionizing wood manufacturing and production*

Industrial revolutions refer to transformative periods in manufacturing and production industries marked by significant technological advancements (see Figure 3; Sherwani *et al.*, 2020). The first industrial revolution, beginning in the 1760s, introduced machinery and mass manufacturing, shifting away from human and animal labor (Haradhan, 2019). The second industrial revolution saw the adoption of electrical engines and oil and gas, while the third industrial revolution, also known as the Technological Revolution, embraced digital information technologies (Dogaru, 2020; Sherwani *et al.*, 2020).

Four Phases of Industrialization

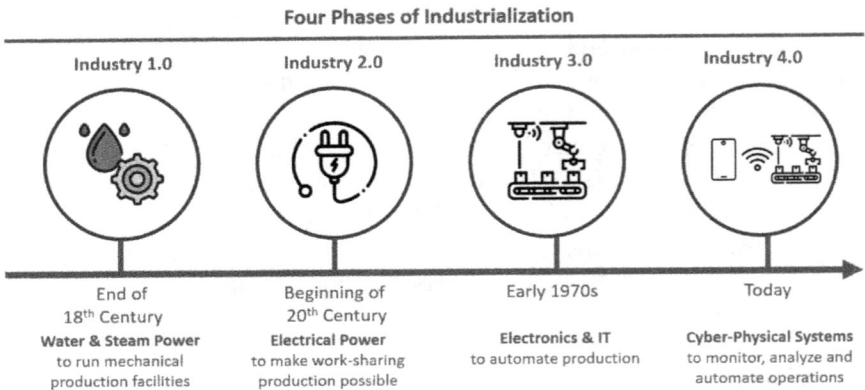

Industry 1.0	Industry 2.0	Industry 3.0	Industry 4.0
End of 18th Century	Beginning of 20th Century	Early 1970s	Today
Water & Steam Power to run mechanical production facilities	Electrical Power to make work-sharing production possible	Electronics & IT to automate production	Cyber-Physical Systems to monitor, analyze and automate operations

Figure 3.   History of industrialization.
Source: Sherwani *et al.* (2020).

The fourth industrial revolution, for which German researchers have coined the moniker Industry 4.0, aims to connect the physical, biological, and digital worlds (Dogaru, 2020). It encompasses various advanced technologies, including cyber-physical systems (CPS), cloud computing, and digital twins (Stock *et al.*, 2018). These are characterized by such attributes as smart networking, mobility, flexibility, customization, and innovation (Jazdi, 2014). CPS links industry's physical components and processes to computer-based algorithms, enabling data-driven operations (Oláh *et al.*, 2020). Cloud computing provides access to shared computing resources via the internet, facilitating scalable and flexible services (Kumar Paul and Ghose, 2012). Digital twins, which create virtual models of physical systems, enabling real-time data collection, sharing, and continuous monitoring. These allow simulating supply chains to test different scenarios and modes for effects of changes to network operations (Zborowski, 2018). Blocker *et al.* (2016) defined Industry 4.0 as the amalgamation of vertical and horizontal supply chains through digitization of products and services, innovative business models, and customer engagement platforms.

Müller *et al.* (2019a) defined three key Industry 4.0 concepts: CPS, internet of things and services (IoTs), and smart factories. CPSs consist of computational and physical abilities that allow human interaction (Baheti and Gill, 2017). They are characterized by high levels of autonomy, giving systems the ability to control physical processes (Boyes *et al.*, 2018).

The IoTs is the network via which CPSs connect (Hermann *et al.*, 2016 as cited in Müller *et al.*, 2019a). It refers to a network of interconnected devices, capable of connecting to the internet (Ande *et al.*, 2020). The concepts of CPS and IoT are coalescing over time to represent hybrid systems consisting of digital, analog, physical and human aspects that function through physics and logic (Greer *et al.*, 2019). The IoTs and data-driven technologies form the basis of smart factories which connect and collect data from various devices, sensors, and systems to enhance productivity, control, and decision making (Elgazzar *et al.*, 2022).

Driving the current industrial revolution are several key technologies, including the IoTs, artificial intelligence (AI), cloud computing, augmented reality, virtual reality, big data analytics, additive manufacturing or 3-D printing, and cybersecurity. These technologies often complement and build upon one another. Advanced robotics, a vital component of Industry 4.0, has revolutionized automation, with collaborative robots enhancing mobility, flexibility, computing capabilities, and programmability (Sherwani *et al.*, 2020). AI, an interconnected system comprising such technologies as big data, machine learning, algorithms, natural language processing, and automation, plays a crucial role in driving intelligent decision-making (Zhang and Lu, 2021). The integration and application of these technologies are revolutionizing the manufacturing and production landscape, enabling increased efficiency, productivity, and innovation across various industries.

2.6. *Need for Industry 4.0 technologies to improve HWP efficiency and carbon storage*

While adoption of Industry 4.0 technologies has been more prevalent in sectors like health, transport, and agriculture, forestry has been slower to embrace these advances (Feng, 2020; Shi *et al.*, 2020). A most recent study indicates that there is potential for these technologies to enhance the assessment and improvement of carbon storage in HWPs (Shivaprakash *et al.*, 2022a).

One challenge observed in the HWP supply chain is a lack of coordination between strategic forest management and tactical industrial production (Kong, 2014; Luo *et al.*, 2021). The former involves long-term aspects such as forest management practices, silvicultural treatments, and market development, while the latter focuses on executing forest management and product manufacturing. Industry 4.0 technologies, including efficient inventory and logistics management systems, can help bridge this gap and improve coordination along the HWP supply chain (Ghobakhloo *et al.*, 2021; Stock *et al.*, 2018). These technologies facilitate collaboration among stakeholders and establish a well-connected industrial network, enabling seamless flow within the supply chain (Ghobakhloo *et al.*, 2021). By integrating these individual systems, Industry 4.0 technologies enhance supply chain management and improve product lifecycle processes (Pödör *et al.*, 2017).

Industry 4.0 technologies can also be incorporated in LCAs in supply chains to monitor and track carbon emissions at each stage, which can provide a better overview of operational improvements to reduce such emissions. Industry technologies such as cyber-physical systems can provide better and more integrated analyses of entire supply chains, from design to end use (Mishra and Singh, 2019). Industry 4.0 technologies also foster sustainable innovation through advanced wood manufacturing techniques, efficient product lifecycle management, sustainable collaboration, and value chain integration (Ghobakhloo *et al.*, 2021). Regarding HWPs, the quality and lifespan of the processed products directly impact the carbon storage potential. A study by Jamwal *et al.* (2021) reveals that leveraging such cutting-edge technologies as additive manufacturing, which enables efficient production and reduces waste through computer-aided design (CAD), can significantly contribute to improved carbon storage in HWPs. Therefore, by utilizing Industry 4.0 technologies, the wood industry can minimize waste, optimize production processes, and produce high-quality products that maximize carbon storage potential.

2.7. *Industry 4.0 technologies in the wood industry*

Digital technologies such as LiDAR, unmanned aerial vehicles (UAV), autonomous vehicles, radio frequency identification (RFID) and remote sensing amongst other technologies have been widely applied in forestry (Feng, 2020), with artificial neural networks (ANNs) being particularly widely applied. Modeled after biological neural systems, ANNs receive input parameters and generate corresponding outputs, eliminating the need to consider such factors as normality and linearity that are present in conventional mathematical models (Eder *et al.*, 2016). ANNs have been extensively used for tree growth models as they are more accurate than traditional statistical regression models (Vieira *et al.*, 2018a).

Important forest management parameters, such as tree height and diameter at breast height (DBH), have been accurately determined using ANNs and other AI tools, including adaptive neuro-fuzzy inference systems (ANFIS) (Bayat *et al.*, 2020; Vieira *et al.*, 2018a). Eder *et al.* (2016) employed ANNs to predict wood volume in a eucalyptus stand.

The earliest application of ANN in the wood industry was to identify varieties, as opposed to manual classification methods (Modasia and De Silva, 2005). ANNs have also been employed to assess bonding strength of wood products and power consumption of wood processing machinery, both crucial for determining product quality (Jovic *et al.*, 2017). The non-linear nature of these processes necessitates the use of ANNs. Such other AI models as support vector machine (SVM), random forest (RF), and deep neural network (DNN) have been utilized in conjunction with satellite images to project such parameters as commercial volume, primary productivity, and stand characteristics (Gonçalves *et al.*, 2021a; Lee *et al.*, 2020a, 2020b; Sakici and Günlü, 2018). Notably, Kam *et al.* (2019) achieved a milestone by 3-D printing wood objects using ink composed entirely of wood, replacing fuel-based resins and utilizing waste.

In small-scale sawmills where data collection and monitoring are limited due to outdated technologies or financial constraints, AI, particularly through machine learning with ANN, has been tested to improve decision-making and prevent technical failures (Borz *et al.*, 2022b; Cheţa *et al.*, 2020b). These studies demonstrated that affordable hardware and software for machine learning enable efficient data processing and monitoring in such sawmills. Real-time tracking of the wood supply chain has also been achieved using Industry 4.0 technology, such as with the German startup Xylene (Shivaprakash *et al.*, 2022a).

Still other forms of AI, such as Deep Learning Algorithms (DLA), have been utilized to assess tree height and diameter dynamics in even-aged pine stands (Ercanlı, 2020) and complex tropical rainforests (Ogana and Ercanli, 2022). DLA models, with their greater complexity and more hidden layers than ANN, exhibit commensurately greater efficiency in characterizing such intricate tropical forest ecosystems (Ercanlı, 2020). AI combined with acoustic signal analysis has been employed for early detection of wood-damaging insect larvae, such as the old house borer. Automated classification using support vector machines (SVM) demonstrated 90 percent accuracy and offered such advantages as lower costs, portability, flexibility, and potentially preventing wood product damage through early detection (Bilski *et al.*, 2017).

Another key Industry 4.0 technology is the aforementioned Internet of Things and services (IoTs), which is additionally described as a network of intelligent objects capable of self-organization, information sharing, and proactive responses to environmental changes (Madakam *et al.*, 2015). In forest management, IoT has found additional applications in such areas as real-time monitoring of logging, forest fire tracking, and continuous vegetation assessment (Singh *et al.*, 2022; Tien Bui *et al.*, 2017). Accounting information systems (AIS) have also proven successful in efficient planning of resource use in wood supply chains by providing an efficient means to link different operational systems to a single well-organized system (Limroscharoen *et al.*, 2018).

Figure 4 illustrates a smart wood supply chain, where data from each link is stored in the cloud, providing access to all stakeholders involved. This interconnected network is referred to as the "internet of trees and services" (Müller *et al.*, 2019b). Remote sensing techniques are employed to create virtual forests, generating valuable data used across the wood supply chain, including the aforementioned harvest planning and operations, transport, and logistics (Müller *et al.*, 2019a). These virtual forests have also been utilized to optimize machinery placement in forests using

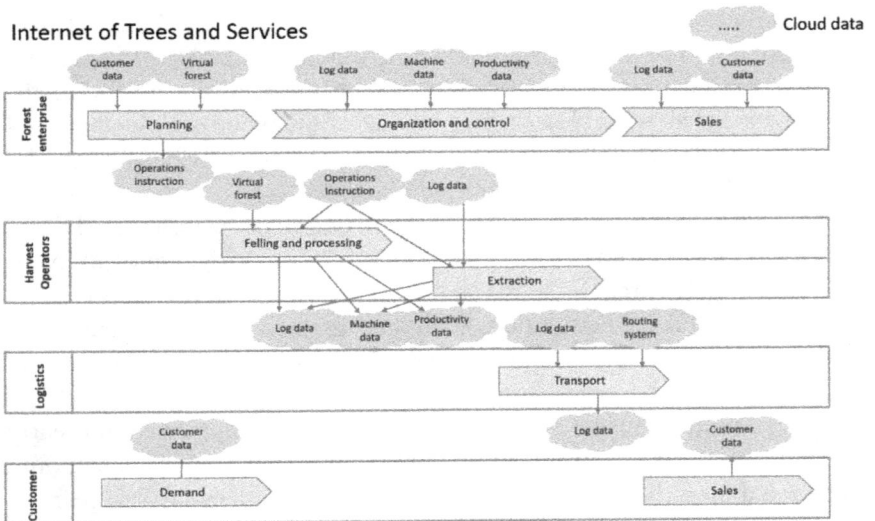

Figure 4.   Wood supply chain within internet of trees and services.

Source: Müller *et al.* (2019b).

the aforementioned satellite imagery, cloud computing, and sensors (Hussein *et al.*, 2015).

2.8. *Industry 4.0 technologies for sustainable development*

In a broader aspect, the application of Industry 4.0 technologies to the SDGs has gained significant attention (Dantas *et al.*, 2020). AI has the potential to contribute to 42 of the 169 SDGs (Hannan *et al.*, 2021). It plays a crucial role in enhancing renewable energy efficiency, operation, and production, reducing emissions and costs. Real-time data collection, monitoring, and inventory management have been made possible through such aforementioned Industry 4.0 technologies as sensing technology, big data, cloud computing, advanced analytics, and AI (Singh *et al.*, 2022).

While Industry 4.0 technologies offer numerous opportunities for sustainability, their impacts can be varied. In terms of manufacturing and industrial processes, these technologies can extend product lifecycles, reduce waste, enhance energy and material efficiency, and facilitate emissions reductions. They enable smart production, additive manufacturing, and data analysis, among other sustainable practices (Ghobakhloo, 2020; Oláh *et al.*, 2020; Stock *et al.*, 2018). They can promote circular economy principles and use of renewables, such as wind and solar energy, which are crucial for sustainable development (He and Ni, 2022; Jamwal *et al.*, 2021). A study by Bag *et al.* (2021) found that Industry 4.0 technologies significantly support the 10 Rs of circular economy. AI, combined with remote sensing, has been widely utilized for real-time monitoring of forest health, deforestation, and illegal logging.

One important sector where Industry 4.0 technologies have significant applications is wood industry supply chains, where they can bring sustainability by greening. The entire operational process of an industry can be optimized and costs reduced through monitoring environmental, social, and economic impacts for better, more timely risk management (Sanders *et al.*, 2019).

Adopting Industry 4.0 technologies, including automation, digitization, and integration, may require new hardware, however, leading to increased energy and materials consumption and higher costs. Older equipment may also become obsolete and discarded in the process, posing environmental challenges (Bonilla *et al.*, 2018; Oláh *et al.*, 2020; Nzetic *et al.*, 2020). Implementing such technologies as IoTs, CPS, and real-time data control necessitates larger data centers, which consume additional energy and investment, with associated potential negative sustainability implications (Bayram, 2020; Bonilla *et al.*, 2018). For example, additive manufacturing allows for efficient prototyping and use of recyclable materials, as well as optimizing carbon storage in HWPs. However, certain materials can consume much more energy than traditional manufacturing (Sanders *et al.*, 2019). Successful adoption of Industry 4.0 technologies can be influenced by such factors as stakeholder doubts about security of cloud storage and other IT systems. This may be especially true of the forest industry, which involves long-term investments and consists of multiple stakeholders scattered across different industries (Bayram, 2020; Gharaibeh *et al.*, 2022). In a broader aspect, careful consideration and planning are crucial to maximizing the benefits of Industry 4.0 technologies on wood industry and sustainability while mitigating potential drawbacks. Integrating these technologies with sustainable practices can thus contribute to achieving SDGs and fostering a more environmentally responsible and resource-efficient future.

3. Discussion and Policy Implications for CCUS

Industry 4.0 technologies have significant implications for CCUS in HWPs, enabling the wood supply chain to adopt more sustainable and efficient practices that enhance carbon storage throughout wood products' lifecycles. ANNs and machine learning algorithms can analyze large datasets, including forest inventories and wood quality parameters, to identify the most suitable wood resources for different product applications. This optimization minimizes waste and maximizes the carbon storage potential of harvested wood products, aligning with CCUS goals (Vieira *et al.*, 2018b).

The real-time process monitoring and control systems that Industry 4.0 affords contribute to improved carbon storage efficiency in wood processing. By integrating sensors, connected devices, and data analytics, these systems enable continuous monitoring of such factors as temperature, humidity, and energy consumption. Timely adjustments and optimizing of processes allow reducing material losses, enhanced product quality, and increased carbon storage (Börz *et al.*, 2022a; Cheţa *et al.*, 2020a). Transparency and traceability are also crucial for ensuring the of wood product sustainability and maximizing carbon storage. Industry 4.0

technologies facilitate collecting, storage, and sharing of data related to wood product origin, processing, and transportation. With blockchain technology, IoT-enabled devices, and digital platforms, stakeholders can track and verify the sustainability credentials of wood products, including responsible sourcing and carbon footprint. This transparency gives incentives to sustainable practices and responsible wood sourcing strategies, leading to enhanced carbon storage in HWPs (Müller *et al.*, 2019a; Shivaprakash *et al.*, 2022b).

With 3-D printing, digital designs, and CAD/CAM, manufacturers can minimize waste, optimize materials usage, and create customized wood products with optimal structural integrity, contributing to long-term carbon storage by enhancing the durability and longevity of finished products (Kam *et al.*, 2019). Industry 4.0 also provides integrated data collection and analytics systems that facilitate lifecycle assessment and product optimization. Analyzing environmental impacts across wood products' lifecycles allows manufacturers to identify opportunities for reducing carbon emissions and enhancing carbon storage. This data-driven approach optimizes product design, material selection, and manufacturing processes, minimizing environmental footprints and maximizing carbon storage potential (Gonçalves *et al.*, 2021b; Lee *et al.*, 2020b).

It can be concluded that Industry 4.0 offers various avenues to enhance carbon storage in HWPs, by enabling efficient resource allocation, real-time process monitoring, traceability, advanced manufacturing techniques, and lifecycle assessment. Embracing these technologies will allow the wood industry to optimize practices, reduce waste, and maximize carbon storage, contributing to CCUS.

3.1. *Policy Support for maximizing HWP carbon storage*

Policymakers play a crucial role in creating an environment enabling fully leveraging the potential of Industry 4.0 in maximizing CCUS along the wood supply chain. The following measures are accordingly recommended: First, policymakers should allocate resources and support research initiatives focused on CCUS (Abidin *et al.*, 2021; Barbieri and Lippert, 2020). This includes encouraging collaboration between CCUS academia, industry partners, and research institutions to develop innovative solutions and enhance understanding of carbon storage mechanisms applicable to the region. Seconds essential for policymakers to establish regulations that offer incentives for adopting Industry 4.0 and promote sustainable practices in the wood industry (Directorate-General for Research and Innovation, 2021; Ramos-Maldonado and Aguilera-Carrasco, 2021), including setting standards for data collection, sharing, and privacy, ensuring reliable and secure use of digital tools for traceability, and certification of sustainably sourced wood products. By creating a clear regulatory framework, policymakers provide guidance that encourages CCUS industry players to embrace these technologies with confidence.

Policymakers should also focus on CCUS capacity building and education (Nielsen *et al.* 2022; Fukushima *et al.*, 2017). Given that Industry 4.0 is relatively new, emphasis should be on equipping stakeholders in the wood industry with the skills and knowledge to effectively utilize these technologies. Equally important, CCUS can provide training programs, workshops, and awareness campaigns specific to local industry, ensuring that individuals have the expertise to leverage Industry 4.0 to drive sustainable practices. Access to adequate financing is also crucial for successful implementation of Industry 4.0 throughout the wood supply chain. Policymakers can therefore support transition to sustainable practices by encouraging green financing initiatives, collaborating with financial institutions (Ramos-Maldonado and Aguilera-Carrasco, 2021), and advocating for favorable loan conditions, subsidies, and tax incentives for businesses adopting these CCUS technologies.

Policymakers should also facilitate public–private partnerships to promote Industry 4.0 in the wood industry, creating platforms for collaboration among industry stakeholders, financial institutions, and government agencies allows streamlining investment procedures and promoting sustainable practices specific to CCUS. Such partnerships can leverage funding opportunities for sustainable wood projects and encourage knowledge sharing. Finally as suggested by Ramos-Maldonado and Aguilera-Carrasco (2021), policymakers should undertake awareness-raising efforts in academia, highlighting potential benefits and the long-term value of investing in Industry 4.0 for the wood industry. Workshops, conferences, and information campaigns can be organized to educate financial institutions about the positive environmental impact and economic viability of sustainable wood practices. Such initiatives can foster increased investment and support for Industry 4.0 applications.

4. Conclusion and Implications

Industry 4.0 adoption holds great promise for maximizing carbon storage in harvested wood products throughout the wood supply chain. Integrating digitalization, automation, and data analytics offers transformative opportunities to enhance sustainability and contribute to CCUS. Stakeholders can thus optimize resource management, minimize waste, and ensure the traceability and sustainability of wood products, maximizing their CCUS potential.

Concerted efforts from various stakeholders are necessary. Policymakers have a crucial role in creating a supportive regulatory framework that encourages implementing CCUS technologies and promotes sustainable practices. They can drive the transition to carbon-efficient wood supply chains and contribute to global climate goals by establishing regulations that offer incentives for adopting Industry 4.0 and CCUS. Collaboration between financial institutions, industry stakeholders, and research institutions is also essential to develop and finance wood industry CCUS projects. Fostering partnerships and creating platforms for collaboration allow stakeholders to leverage funding opportunities, streamline investment procedures, and facilitate Industry 4.0 wood sector implementation for CCUS.

Looking ahead, it is crucial to continue Industry 4.0 research and development for the wood industry with emphasis on CCUS. This includes exploring innovative solutions, enhancing the accuracy of predictive models, and further integrating remote sensing and AI to optimize CCUS. It can be concluded that Industry 4.0 holds great promise for contributing to CCUS, maximizing carbon storage in HWPs throughout the wood supply chain. Embracing these technologies, implementing supportive policies, fostering collaborations, and advancing research allow creating a sustainable and resilient wood industry that actively contributes to CCUS, playing a vital role in mitigating climate change.

References

Baheti, R. and Gill, H. (2017). Cyber-physical systems. *The Impact of Control Technology*, *12*, 161–166. https://doi.org/10.1007/s10559-017-9984-9.

Barua, S. K., Lehtonen, P., and Pahkasalo, T. (2014). Plantation vision: Potentials, challenges and policy options for global industrial forest plantation development. *International Forestry Review*, *16*, 117–127.

Bayat, M., Vieira, R., and Amiri, B. (2020). Application of artificial neural network for estimating tree diameter and height in mixed uneven-aged forests. *Forest Ecology and Management, 460*, 117876.

Bayram, B. C. (2020). The impact of Industry 4.0 on supply chains. In *Conference International Forestry & Nature Tourism E-Congress*.

Bilski, P., Bobiński, P., Krajewski, A., and Witomski, P. (2017). Detection of wood boring insects' larvae based on the acoustic signal analysis and the artificial intelligence algorithm. *Archives of Acoustics, 42*, 61–70. https://doi.org/10.1515/aoa-2017-0007.

Blocker, M., Mundoch, I., Bromley, K., Geissbauer, R., Vedso, J., and Schrauf, S. (2016). Industry 4.0: Building the digital enterprise: Forest, paper and packaging key findings. In *2016 Global Industry 4.0 Survey: Industry Key Findings, 73 Forest, Paper and Packaging Company Executives Interviewed in 26 Countries*.

Bont, L. G., Fraefel, M., Frutig, F., Holm, S., Ginzler, C., and Fischer, C. (2022). Improving forest management by implementing best suitable timber harvesting methods. *Journal of Environmental Management, 302*, 114099. https://doi.org/10.1016/j.jenvman.2021.114099.

Börz, S. A., Borz, S. A., Dănilă, I., Iancu, L., and Mihalache, G. (2022a). Efficiency analysis of small-scale sawmills using artificial intelligence techniques. *Journal of Cleaner Production, 334*, 147359.

Borz, S. A., Forkuo, G. O., Oprea-Sorescu, O., and Proto, A. R. (2022b). Development of a robust machine learning model to monitor the operational performance of fixed-post multi-blade vertical sawing machines. *Forests, 13*, 1–24. https://doi.org/10.3390/f13071115.

Boyes, H., Hallaq, B., Cunningham, J., and Watson, T. (2018). The industrial internet of things (IIoT): An analysis framework. Comput. Ind. *101*, 1–12. https://doi.org/10.1016/j.compind.2018.04.015.

Brunet-Navarro, P., Jochheim, H., Kroiher, F., and Muys, B. (2018). Effect of cascade use on the carbon balance of the German and European wood sectors. *Journal of Cleaner Production, 170*, 137–146. https://doi.org/10.1016/j.jclepro.2017.09.135.

Brunet-Navarro, P., Jochheim, H., and Muys, B. (2017). The effect of increasing lifespan and recycling rate on carbon storage in wood products from theoretical model to application for the European wood sector. *Mitigation and Adaptation Strategies for Global Change, 22*, 1193–1205. https://doi.org/10.1007/s11027-016-9722-z.

Cheța, I., Ionescu, S., and Briciu, A. V. (2020a). Integration of a wood waste monitoring system in the concept of Industry 4.0. *Sustainability, 12*(3), 1166.

Cheța, M., Marcu, M.V., Iordache, E., and Borz, S.A. (2020b). Testing the capability of low-cost tools and artificial intelligence techniques to automatically detect operations done by a small-sized manually driven bandsaw. *Forests, 11*, 739. https://doi.org/10.3390/F11070739.

Dalalah, D., Khan, S. A., Al-Ashram, Y., Albeetar, S., Ali, Y. A., Alkhouli, E. (2022). An integrated framework for the assessment of environmental sustainability in wood supply chains. *Environmental Technology & Innovation*, *27*, 102429. https://doi.org/10.1016/j.eti.2022.102429.

Dantas, T. E. T., De-Souza, E. D., Destro, I. R., Hammes, G., Rodriguez, C. M. T., Soares, S. R. (2020). How the combination of Circular Economy and Industry 4.0 can contribute towards achieving the sustainable development goals. *Sustainable Production and Consumption*. 26. https://doi.org/10.1016/j.spc.2020.10.005.

Directorate-General for Research and Innovation (2021). Industry 5.0: Towards more sustainable, resilient and human-centric industry. European Commission. https://research-and-innovation.ec.europa.eu.

Eder, P. M., FabrÃcia, C. M. M., Saulo, J. T., Rodrigo, G. M. N., FabrÃcio, A. L., Reginaldo, S., P., Alba, V. R. (2016). Artificial intelligence tools in predicting the volume of trees within a forest stand. *African Journal of Agricultural Research*, *11*, 1914–1923. https://doi.org/10.5897/ajar2016.11015.

Elgazzar, K., Khalil, H., Alghamdi, T., Badr, A., Abdelkader, G., Elewah, A., and Buyya, R. (2022). Revisiting the internet of things: New trends, opportunities and grand challenges. *Frontiers in the Internet of Things*, *1*, 1–18. https://doi.org/10.3389/friot.2022.1073780.

Ercanlı, İ. (2020). Innovative deep learning artificial intelligence applications for predicting relationships between individual tree height and diameter at breast height. *Forest Ecosystems*. 7. https://doi.org/10.1186/s40663-020-00226-3.

FAO. (2020a). Global forest resources assessment, key findings. https://www.fao.org/3/CA8753EN/CA8753EN.pdf.

FAO. (2020b). Natural forest management [WWW document]. https://www.fao.org/forestry/sfm/85084/en/.

FAO. (Food and Agriculture Organization of the United Nations). (2022). Climate change mitigation and harvested wood products: Lessons learned from three case studies in Asia and the Pacific.

Feng, Y. (2020). Forestry 4.0: A framework for the forest supply chain toward Industry 4.0. *Gestão & Produção*, *27*.

Fukushima, Y., Ishimura, G., Komasinski, A. J., Omoto, R., and Managi, S. (2017). Education and capacity building with research: A possible case for future Earth. *International Journal of Sustainability in Higher Education*, *18*(2), 263–276.

Ghaffariyan, M. R. (2021). Review of studies on motor-manual felling productivity in eucalypt stands. *Silva Balcanica*, *22*, 77–87. https://doi.org/10.3897/silvabalcanica.22.e58750.

Gharaibeh, L., Eriksson, K., and Lantz, B. (2022). Supply chain digitalization in the wood manufacturing industry: A bibliometric literature review. *Advanced Transdisciplinary Engineering and Technology*, *21*, 617–628. https://doi.org/10.3233/ATDE220180.

Gonçalves, A. C., Silva, C. A., and Tomé, M. (2021b). Prediction of Eucalyptus wood density using random forest algorithm and remote sensing data. *Remote Sensing, 13*(8), 1575.

Gonçalves, F. C., Miguel, E. P., Matricardi, E. A. T., Emmert, F., and Santana, C. C. (2021a). Artificial intelligence associated with Sentinel-2 data in predicting commercial volume in Brazilian Amazon Forest. *Journal of Applied Remote Sensing, 15*, 1–19. https://doi.org/10.1117/1.jrs.15.044511.

Greer, C., Burns, M., Wollman, D., and Griffor, E. (2019). Cyber-physical systems and internet of things [WWW document]. Gaithersburg, MD.

Haddouche, M. and Ilinca, A. (2022). Energy efficiency and industry 4.0 in wood industry: A review and comparison to other industries. *Energies, 15*, 2384. https://doi.org/10.3390/en15072384.

Hannan, M. A., Al-Shetwi, A. Q., Ker, P. J., Begum, R. A., Mansor, M., Rahman, S. A., Dong, Z. Y., Tiong, S. K., Mahlia, T. M. I., Muttaqi, K. M. (2021). Impact of renewable energy utilization and artificial intelligence in achieving sustainable development goals. *Energy Reports, 7*, 5359–5373. https://doi.org/10.1016/j.egyr.2021.08.172.

He, Z. and Turner, P. (2021). A systematic review on technologies and Industry 4.0 in the forest supply chain: A framework identifying challenges and opportunities. *Logistics, 5*. https://doi.org/10.3390/logistics5040088.

Hussein, M., Renner, M., and Iagnemma, K. (2015). Global localization of autonomous robots in forest environments. *Photogrammetric Engineering and Remote Sensing, 81*, 839–846. https://doi.org/10.14358/PERS.81.11.839.

Iordan, C. M., Hu, X., Arvesen, A., Kauppi, P., and Cherubini, F. (2018). Contribution of forest wood products to negative emissions: Historical comparative analysis from 1960 to 2015 in Norway, Sweden and Finland. *Carbon Balance and Management, 13*. https://doi.org/10.1186/s13021-018-0101-9.

IPCC. (2006). *2006 IPCC Guidelines for National Greenhouse Gas Inventories.* Chapter 12 Harvested Wood Products, Yokohama, Japan. https://www.ipcc-nggip.iges.or.jp/public/2006gl/pdf/4_Volume4/V4_12_Ch12_HWP.pdf.

IPCC. (2014). 2013 Revised Supplementary Methods and Good Practice Guidance Arising from the Kyoto Protocol.

IRENA. (2009). *Bioenergy from Boreal Forests. Swedish Approach to Sustainable Wood Use.* IRENA, Svebio and Swedish Energy Agency, 2019.

Jazdi, N. (2014). Cyber physical systems in the context of Industry 4.0. *Proc. 2014 IEEE International Conference Automation, Quality and Testing, Robotics AQTR 2014*, 0–4. https://doi.org/10.1109/AQTR.2014.6857843.

Johnston, C. M. T. and Radeloff, V. C. (2019). Global mitigation potential of carbon stored in harvested wood products. *Proceedings of the National Academy of Sciences of the United States of America, 116*, 14526–14531. https://doi.org/10.1073/pnas.1904231116.

Jonsson R. *et al.* (2021). Boosting the EU forest-based bioeconomy: Market, climate, and employment impacts. *Technological Forecasting and Social Change*, 163, 120478.

Jourgholami, M., Majnounian, B., and Zargham, N. (2013). Performance, capability and costs of motor-manual tree felling in hyrcanian hardwood forest. *Croatian Journal of Forest Engineering*, *34*, 283–293.

Jovic, S., Golubovic, Z., and Stojanovic, J. (2017). Wood bonding strength sensitivity estimation and power consumption prediction in wood machining process by artificial intelligence methods. *Sensor Review*, *37*, 444–447. https://doi.org/10.1108/SR-06-2017-0119

Jürgensen, C., Kollert, W., and Lebedys, A. (2014). Assessment of industrial roundwood production from planted forests. FAO Plant. For. Trees Work. Pap. FP/48/E, 40.

Kaakkurivaara, N. and Stampfer, K. (2018). Assessment for improvement: Harvesting operations in small-scale forest on Thai steep terrain. *Small-scale Forestry*, *17*, 259–276. https://doi.org/10.1007/s11842-017-9386-x.

Kaarakka, L., Cornett, M., Domke, G., Ontl, T., and Dee, L. E. (2021). Improved forest management as a natural climate solution: A review. *Ecological Solutions and Evidence*, *2*, 1–10. https://doi.org/10.1002/2688-8319. 12090.

Kam, D., Gardner, D., Hall, M., and Belov, I. (2019). Additive manufacturing with wood-based filaments. *Additive Manufacturing*, *26*, 73–78.

Kong, J. (2014). Coordination between strategic forest management and tactical logistic and production planning in the forestry supply chain. *International Transactions in Operational Research*, *21*, 1–33. https://doi.org/10.1111/itor.12089.

Lee, J. H., Singh, R. M., Lee, W. K., and Kim, J. (2020a). Analysis of factors influencing the primary productivity of forests using remote sensing and machine learning techniques. *Remote Sensing*, *12*(12), 1934.

Lee, B., Kim, N., Kim, E. S., Jang, K., Kang, M., Lim, J. H., Cho, J., and Lee, Y. (2020b). Primary productivity in the forests of South Korea using satellite remote sensing data. *Forests*, 11. https://doi.org/10.3390/f11091000.

Lee, N., Cardoso de Oliveira, R., Roberts, B., Katz, J., Brown, T., and Flores-Espino, F. (2020c). Exploring renewable energy opportunities in select Southeast Asian countries: A geospatial analysis of the levelized cost of energy of utility-scale wind and solar photovoltaics. USAID-NREL Partnership, Contract No. IAG-17-2050.

Legg, *et al.* (2021). Industry 4.0 Implementation in US Primary Wood Products Industry. *Drvna Industrija*, *72*(2), 143–153. https://doi.org/10.5552/drvind. 2021.2017.

Ligot, G., Fayolle, A., Gourlet-fleury, S., Dainou, K., Gillet, J., Ridder, M. De, Drouet, T., Groenendijk, P., and Doucet, J. (2019). Forest ecology and management growth determinants of timber species Triplochiton scleroxylon and implications for forest management in central Africa. *Forest Ecology and Management*, *437*, 211–221. https://doi.org/10.1016/j.foreco.2019.01.042.

Limroscharoen, S., Duangphakdee, A., and Suttipun, M. (2018). What are the key factors influencing successful implementation of accounting information systems for resource planning? A case study of rubber wood companies in Thailand. *ABAC ODI Journal. Vision Action Outcome*, *5*, 130–143.

Liu, G. and Han, S. (2009). Long-term forest management and timely transfer of carbon into wood products help reduce atmospheric carbon. *Ecological Modelling*, *220*, 1719–1723. https://doi.org/10.1016/j.ecolmodel.2009.04.005.

Luo, L., Hehir, J. O., Regan, C. M., Meng, L., Connor, J. D., Chow, C. W. K. (2021). Forest Policy and Economics: An integrated strategic and tactical optimization model for forest supply chain planning. *Forest Policy and Economics*, *131*, 102571. https://doi.org/10.1016/j.forpol.2021.102571.

Madakam, S., Ramaswamy, R., and Tripathi, S. (2015). Internet of Things (IoT): A literature review. *Journal of Computing and Communications*, *3*, 164–173. https://doi.org/10.4236/jcc.2015.35021.

Manavakun, N. (2014). Harvesting operations in eucalyptus plantations in Thailand. *Dissertationes Forestales*, *177*, 111. https://doi.org/10.14214/df.177.

Mcewan, A., Marchi, E., Spinelli, R., and Brink, M. (2020). Past, present and future of industrial plantation forestry and implication on future timber harvesting technology. *Journal of Forestry Research*, *31*, 339–351. https://doi.org/10.1007/s11676-019-01019-3.

Meyer, R., Campanella, S., Corsano, G., and Montagna, J. M. (2019). Optimal design of a forest supply chain in Argentina considering economic and social aspects. *Journal of Cleaner Production*, *231*, 224–239. https://doi.org/10.1016/j.jclepro.2019.05.090.

Michael, J. and Gorucu, S. (2021). Occupational tree felling fatalities: 2010–2020. *American Journal of Industrial Medicine*, *64*, 969–977. https://doi.org/10.1002/ajim.23286.

Mishra, A., Humpenöder, F., Dietrich, J. P., Bodirsky, B. L., Sohngen, B., Reyer, C. P. O., Lotze-Campen, H., and Popp, A. (2021). Estimating global land system impacts of timber plantations using MAgPIE 4.3.5. *Geoscientific Model Development*, *14*, 6467–6494. https://doi.org/10.5194/gmd-14-6467-2021.

Mishra, S. and Singh, S. P. (2019). Carbon management framework for sustainable manufacturing using life cycle assessment. *IoT and Carbon Sequestration*, *28*, 1396–1409. https://doi.org/10.1108/BIJ-01-2019-0044.

Modasia, B. and De Silva, M. A. (2005). An intelligent system to classify varieties of wood, *Sri Lanka Association for Artificial Intelligence*, 61–69.

Müller, D., Stoffel, M., Fontana, V., and Grêt-Regamey, A. (2019a). Digital transformation in the forest-based sector: A systematic literature review on impacts and implications. *Forest Policy and Economics*, *109*, 102021.

Müller, F., Jaeger, D., and Hanewinkel, M. (2019b). Digitization in wood supply — A review on how Industry 4.0 will change the forest value chain. *Computer and Electronics in Agriculture*, *162*, 206–218. https://doi.org/10.1016/j.compag.2019.04.002.

Nielsen, J. A. E., Stavrianakis, K., and Morrison, Z. (2022). Community acceptance and social impacts of carbon capture, utilization and storage projects: A systematic meta-narrative literature review. *PLoS ONE*, *17*(8), e0272409.

Nepal, P., Grala, R. K., and Grebner, D. L. (2012). Financial feasibility of increasing carbon sequestration in harvested wood products in Mississippi. *Forest Policy and Economics*, *14*, 99–106. https://doi.org/10.1016/j.forpol.2011.08.005.

Nepal, P., Johnston, C. M. T., and Ganguly, I. (2021). Effects on global forests and wood product markets of increased demand for mass timber. *Sustainability*, *13*. https://doi.org/10.3390/su132413943.

Nepal, P., Korhonen, J., Prestemon, J. P., and Cubbage, F. W. (2019). Projecting global planted forest area developments and the associated impacts on global forest product markets. *Journal of Environmental Management*, *240*, 421–430. https://doi.org/10.1016/j.jenvman.2019.03.126.

Nzetic, S., Solic, P., Lopez-de-Ipina, Gonzalez-de-Artaza, D., and Patrono, L. (2020). Internet of Things (IoT): Opportunities, issues and challenges towards a smart and sustainable future. *Journal of Cleaner Production*, *6*, 1–33.

Ogana, F. N. and Ercanli, I. (2022). Modelling height-diameter relationships in complex tropical rain forest ecosystems using deep learning algorithm. *Journal of Forestry Research*, *33*, 883–898. https://doi.org/10.1007/s11676-021-01373-1.

Ontl, T. A., Janowiak, M. K., Swanston, C. W., Daley, J., Handler, S., Cornett, M., Hagenbuch, S., Handrick, C., McCarthy, L., and Patch, N. (2020). Forest management for carbon sequestration and climate adaptation. *Journal of Forestry*, *118*, 86–101. https://doi.org/10.1093/jofore/fvz062.

Paluš, H., Parobek, J., Moravčík, M., Kovalčík, M., Dzian, M., and Murgaš, V. (2020). Projecting climate change potential of harvested wood products under different scenarios of wood production and utilization: Study of Slovakia. *Sustainability*, *12*. https://doi.org/10.3390/su12062510.

Parobek, J., Paluš, H., Moravčík, M., Kovalčík, M., Dzian, M., Murgaš, V., and Šimo-Svrček, S. (2019). Changes in carbon balance of harvested wood products resulting from different wood utilization scenarios. *Forests*, *10*. https://doi.org/10.3390/f10070590.

Peña-Claros, M., Fredericksen, T. S., Alarcón, A., Blate, G.M., Choque, U., Leaño, C., Licona, J. C., Mostacedo, B., Pariona, W., Villegas, Z., and Putz, F. E. (2008). Beyond reduced-impact logging: Silvicultural treatments to increase growth rates of tropical trees. *Forest Ecology and Management*, *256*, 1458–1467. https://doi.org/10.1016/j.foreco.2007.11.013.

Pödör, Z., Gludovátz, A., Bacsárdi, L., Erdei, I., and Janky, F. N. (2017). Industrial IoT techniques and solutions in wood industrial manufactures. *Infocommunications Journal*, *9*, 24–30.

Profft, I., Mund, M., Weber, G. E., Weller, E., Schulze, E. D. (2009). Forest management and carbon sequestration in wood products. *European Journal of Forest Research*, *128*, 399–413. https://doi.org/10.1007/s10342-009-0283-5.

Ramos-Maldonado, M. and Aguilera-Carrasco, C. (2021). Trends and opportunities of Industry 4.0 in wood manufacturing processes. In Gong, M. (ed.), *Engineered Wood Products for Construction*. IntechOpen, 1–14. New Brunswick, Canada.

Sakici, O. E. and Günlü, A. (2018). Artificial intelligence applications for predicting some stand attributes using landsat 8 oli satellite data: A case study from Turkey. *Applied Ecology and Environmental Research*, *16*, 5269–5285. https://doi.org/10.15666/aeer/1604_52695285.

Sanders, N. R., Boone, T., Ganeshan, R., Wood, J. D. (2019). Sustainable supply chains in the age of AI and digitization: Research challenges and opportunities. *Journal of Business Logistics*, *40*, 229–240. https://doi.org/10.1111/jbl.12224.

Santos, A., Carvalho, A., Barbosa-Póvoa, A. P., Marques, A., and Amorim, P. (2019). Assessment and optimization of sustainable forest wood supply chains — A systematic literature review. *Forest Policy and Economics*, *105*, 112–135. https://doi.org/10.1016/j.forpol.2019.05.026.

Sasaki, N. (2021). Timber production and carbon emission reductions through improved forest management and substitution of fossil fuels with wood biomass. *Resources, Conservation and Recycling*, *173*, 105737. https://doi.org/10.1016/j.resconrec.2021.105737.

Sasaki, N., Knorr, W., Foster, D. R., Etoh, H., Ninomiya, H., Chay, S., Kim, S., Sun, S. (2009). Woody biomass and bioenergy potentials in Southeast Asia between 1990 and 2020. *Applied Energy*, *86*, S140–S150.

Sasaki, N., Owari, T., Putz, F. E. (2011). Time to substitute wood bioenergy for nuclear power in Japan. *Energies*, *4*(7), 1051–1057.

Sasaki, N., Chheng, K., and Ty, S. (2012). Managing production forests for timber production and carbon emission reductions under the REDD+ scheme. *Environmental Science & Policy*, *23*, 35–44.

Sasaki, N., Asner, G. P., Pan, Y., Knorr, W., Durst, P. B., Ma, H. O., Abe, I., Lowe, A. J., Koh, L. P., and Putz, F. E. (2016). Sustainable management of tropical forests can reduce carbon emissions and stabilize timber production. *Frontiers in Environmental Science*, *4*, 50. doi: 10.3389/fenvs.2016.00050.

Sasaki, N., Myint, Y. Y., Abe, I., and Venkatappa, M. (2021). Predicting carbon emissions, emissions reductions, and carbon removal due to deforestation and plantation forests in Southeast Asia. *Journal of Cleaner Production, 312*, 127728. https://doi.org/10.1016/j.jclepro.2021.127728.

Sherwani, F., Asad, M. M., Ibrahim, B. S. K. K. (2020). Collaborative robots and Industrial Revolution 4.0 (IR 4.0). *2020 Int. Conf. Emerg. Trends Smart Technol. ICETST 2020 0.* https://doi.org/10.1109/ICETST49965.2020.9080724.

Shi, Z. R., Wang, C., and Fang, F. (2020). Artificial Intelligence for Social Good: A Survey, 1–78.

Shivaprakash, K. N., Swami, N., Mysorekar, S., Arora, R., Gangadharan, A., Vohra, K., Jadeyegowda, M., and Kiesecker, J. M. (2022a). Potential for artificial intelligence (AI) and machine learning (ML) applications in biodiversity conservation, managing forests, and related services in India. *Sustainability, 14*, 1–20. https://doi.org/10.3390/su14127154.

Shivaprakash, N. C., Gopalappa, T., and Shivaprasad, N. (2022b). Blockchain-based supply chain traceability for sustainable wood products. *Journal of Cleaner Production*, 339, 129849.

Singh, R., Gehlot, A., Vaseem Akram, S., Kumar Thakur, A., Buddhi, D., and Kumar Das, P. (2022). Forest 4.0: Digitalization of forest using the Internet of Things (IoT). *Journal of King Saud University — Computer and Information Sciences, 34*, 5587–5601. https://doi.org/10.1016/j.jksuci.2021.02.009.

Stantutf, J. A., Kellison, R. C., Broerman, F. S., and Jones, S. B. (2003). Productivity of southern pine plantations. Where are we and how did we get here? *Journal of Forestry*, 101.

Tian, X., Sohngen, B., Kim, J.B., Ohrel, S., and Cole, J. (2016). Global climate change impacts on forests and markets. *Environmental Research Letters, 11*, 035011.

Tien Bui, D., Bui, Q. T., Nguyen, Q. P., Pradhan, B., Nampak, H., and Trinh, P. T. (2017). A hybrid artificial intelligence approach using GIS-based neural-fuzzy inference system and particle swarm optimization for forest fire susceptibility modeling at a tropical area. *Agricultural and Forest Meteorology, 233*, 32–44. https://doi.org/10.1016/j.agrformet.2016.11.002.

United Nations Economic and Social Commission for Asia and the Pacific (2022). Asia and the Pacific SDG Progress Report 2022. ESCAP. https://www.unescap.org.

Vass, M. M. and Elofsson, K. (2016). Is forest carbon sequestration at the expense of bioenergy and forest products cost-efficient in EU climate policy to 2050? *Journal of Forest Economics*, 24, 82–105.

Vieira, G. C., de Mendonça, A. R., da Silva, G. F., Zanetti, S. S., da Silva, M. M., and dos Santos, A. R. (2018a). Prognoses of diameter and height of trees of eucalyptus using artificial intelligence. *Science of the Total Environment, 619–620*, 1473–1481. https://doi.org/10.1016/j.scitotenv.2017.11.138.

Vieira, R., Bayat, M., Oliveira, T., and Tomé, M. (2018b). Estimating tree diameter and height in mixed uneven-aged forests using artificial neural networks. *Ecological Indicators*, *94*, 235–242.

Werner, F., Taverna, R., Hofer, P., Thürig, E., and Kaufmann, E. (2010). National and global greenhouse gas dynamics of different forest management and wood use scenarios: A model-based assessment. *Environmental Science & Policy*, *13*, 72–85. https://doi.org/10.1016/j.envsci.2009.10.004.

Part 3

Chapter 8

Carbon Pricing and CCUS: Evidence from China

Anupam Dutta[*,‖], Mohammad Rakib Uddin Bhuiyan[†,**],
Gang-Jin Wang[‡,††], Gazi Salah Uddin[§,‡‡], and Ali Ahmed[¶,§§]

[*]*School of Accounting and Finance, University of VAASA,
Vaasa, Finland*

[†]*Division of Economics, Linköping University, Linköping, Sweden
Associate Professor (on leave), University of Dhaka, Bangladesh*

[‡]*Business School, Hunan University, Changsha, P. R. China*

[§]*Division of Economics, Department of Management and Engineering,
Linköping University, Linköping, Sweden*

[¶]*Division of Economics, Department of Management and Engineering,
Linköping University, Linköping, Sweden
The Ratio Institute, Stockholm, Sweden*

[‖]*adutta@uwasa.fi*

[**]*Mohammad.rakib.bhuiyan@liu.se*

[††]*wanggangjin@hnu.edu.cn*

[‡‡]*gazi.salah.uddin@liu.se*

[§§]*ali.ahmed@liu.se*

Abstract

While the process of carbon capture, utilization, and storage (CCUS) plays a pivotal role in mitigating climate change impacts, rising economic uncertainty, geopolitical conflict, and oil price volatility tend to retard CCUS deployment; which carbon emissions trading mechanisms can mitigate. The literature shows that such schemes are still immature in developing economies such as China, where carbon pricing seems to be a key strategy to lower CO_2 power generation emissions. In this study, we thus investigate the Chinese carbon market's volatility, concentrating on time-dependent jumps in emissions pricing. As jump-induced volatility represents an important risk, precise information thereon is important for increased carbon trading efficiency. The GARCH-jump process finds that such jumps do occur in the Chinese emissions market and that key uncertainty indicators including the aforementioned economic policy uncertainty, crude oil volatility index, and geopolitical risk can explain the resulting volatility, with important implications for policymakers and socially responsible investors.

Keywords: CCUS, carbon trading, climate change, China, time-varying jumps.

1. Introduction

Carbon capture, utilization, and storage (CCUS) plays a pivotal role in mitigating climate change impacts. Its objectives include managing emissions from heavy industry and removing carbon from the atmosphere, in which it is considered an efficient means to achieve net-zero targets. Although CCUS has lately received attention from governments, policymakers, and investors, its deployment lags. An International Energy Agency (IEA) report reveals that promoting CCUS accounts for only 0.5 percent of global investment in generating renewable and sustainable energy. The COVID-19 pandemic and the resulting economic downturn and low oil prices are also obstacles to CCUS deployment.

Carbon emissions trading mechanisms can and may overcome handicaps to CCUS project developments (Lin and Tan, 2021). Given that, as noted earlier, the current economic environment may be insufficient to prompt rapid CCUS investments; an upgraded emissions trading system can make projects more attractive, while carbon pricing offers policymakers the flexibility to determine the cheapest ways to reduce carbon emissions. Emission trading schemes remain immature in developing economies such as China, however, where carbon pricing seems to be a key strategy to reduce CO_2 power generation emissions, while CCUS technologies are inadequate for reaching carbon neutrality. The Chinese government must make carbon trading markets more efficient as this will help drive CCUS deployment. Such markets may be highly volatile, being a new financial asset class (Weng and Xu, 2018), necessitating precise estimates of time-varying volatilities if their efficiencies are to be quantified. Qi *et al.* (2022) argue that accurate volatility forecasts are essential in Chinese carbon markets if investors and regulators are to make proper decisions, which would further aid in reducing greenhouse gas emissions (GHGs). Additionally, Lin and Tan (2021) shed light on the importance of improved, more efficient Chinese emissions trading markets, which would be a primary CCUS development strategy.

Given the significance of carbon trading to CCUS deployment, a growing body of literature investigates these Chinese carbon emissions market volatility dynamics. Wang and Guo (2018) examine the volatility linkage between carbon and energy markets, whereas Xu (2021) explores how energy market uncertainties affect Chinese emissions pricing volatility. Chun (2018) estimates volatility cross-effects between Chinese and EU emissions markets. Other notable works include Ji *et al.* (2019), Lin and Chen (2019), and Zhu *et al.* (2020) who also explain the phenomena.

While more studies focus on the Chinese emissions trading scheme, none has investigated the jump behavior of carbon pricing. Time-varying jumps, which are often seen in such immature markets, may drive volatility substantially higher. This represents an important risk, demanding particular attention when modeling carbon market volatility.[1] The main objective of this study is to examine whether time-varying jumps occur in Chinese emissions prices. We also assess whether relevant variables, including the aforementioned oil price risk, economic policy uncertainty, and geopolitical risk (GPR), may explain such jump-induced volatility (JV). Identifying factors influencing carbon price volatility will help policymakers precisely estimate the underlying risk linked to carbon emissions trading. This strand of research also has important implications for CCUS development in China, which, as mentioned previously, is essential to achieving net-zero targets.

Methodologically, we employ the GARCH-jump model proposed by Chan and Maheu (2002), finding that time-dependent jumps do occur in

---

[1] Dutta (2018) shows that the EU emission market also experiences such jumps, which play a pivotal role in volatility modeling and risk management.

the Chinese emissions market, and that the aforementioned key uncertainty indicators, including economic policy uncertainty (EPU), the crude oil volatility index (OVX), and geopolitical risk (GPR), can indeed explain the Chinese JV. Given that precise JV information is crucial to greater carbon emissions market efficiency, our analysis offers important implications to policymakers and socially responsible investors alike.

2. Literature Review

China, the largest emitter of CO_2 in the world, faces ethical and normative pressure to adopt sustainable energy development policies and reduce its GHGs. At the UN General Assembly's 75th session, President Xi Jinping announced that China is dedicated to achieving peak CO_2 emissions by 2030 and carbon neutrality by 2060. He also urges the global community to pursue innovative, coordinated, green, and open development for the betterment of humanity. As part of this vision, carbon pricing can substantially reduce GHGs (Mo *et al.*, 2023; Yan and Yang, 2021; Zhao *et al.*, 2022). China's carbon markets are evolving with commodity and financial attribution (Wen *et al.*, 2022) and regulating the most significant GHGs (Liao *et al.*, 2022). The pricing mechanisms in these markets chiefly comprise emissions trading systems (ETS), a.k.a. cap and trade, carbon taxes, and a hybrid of these (Narassimhan *et al.*, 2018).

Carbon pricing in the form of a carbon tax provides a double dividend, improving economic efficiency by redistributing collected, carbon tax revenue and reducing GHGs (Kojima and Asakawa, 2021). There is extensive literature (Compernolle *et al.*, 2022; Landis *et al.*, 2021; Li *et al.*, 2018; Mo *et al.*, 2023; van den Bergh and Savin, 2021; Venmans *et al.*, 2020; Wen *et al.*, 2022) on carbon pricing in various contexts, as well as traditional and emerging carbon economics concerns. Lin and Jia (2019) explore the role of both carbon trading and carbon taxes on China's environment, energy, and economy, concluding that, while both can substantially reduce GHGs, the tax is more efficient than carbon trading. They also point out the negative effect of carbon trading on the output of the energy industry and energy-intensive industry alike, including implicit costs of setting up a carbon trading market. Thus, they suggest levying a carbon tax on fossil fuels in China to reduce CO_2 emissions effectively. Mo *et al.* (2023) show that China's carbon emission trading scheme (ETS) with carbon price stabilization mechanism (CPSM) has varying impact on Chinese firms' productivity. An ETS with explicit price corridors could improve total factor productivity (TFP) by 0.381, whereas an ETS without such corridors could influence TFP by 0.198, while ETS without CPSM reduces TFP by 0.352. Extant works (Känzig, 2023; Ko and Lee, 2022; Wen *et al.*, 2022) reveal nonlinear asymmetric and unstable effects of carbon prices on economic agents, urging equitable carbon policy and market development.

Carbon market development is credited with affirmation of carbon pricing, which has a decisive role in carbon trading as it strengthens resource distribution and optimization (Fang *et al.*, 2018). It has thus been in use as a policy instrument for climate mitigation in many countries, although its application varies by context. Political systems, domestic business, competition, and influence over local authorities, including international obligations, contribute significantly to market design, adoption, and implementation (Khan and Johansson, 2022). In 2005, the European Union (EU) launched the first carbon market, laying the foundation for emerging policy-based markets worldwide. China followed in 2013 with seven carbon pilot markets, which it replaced with a national market on July 16, 2021. With energy-intensive heavy industry, China has significant potential for operating a robust carbon market. A gap exists between the mature carbon market in the West and the infant carbon market in China, however, because of differences in such contextual factors as carbon price fixation, quota allocation, and emissions coverage sources (Hua and Dong, 2019).

The differences between the EU ETS and China's national ETS are complementary, and linking these systems may offer economic and political advantages (Zeng *et al.*, 2016). International cooperation on carbon pricing through linking carbon markets can deliver economic benefits by lowering carbon reduction costs, and environmental benefits by reducing carbon emissions and leakage (Thube *et al.*, 2021). As mentioned above, China's national carbon market is in its early stages and requires additions and improvements. The most successful EU carbon market can accordingly provide valuable lessons for China's national carbon market (Liu *et al.*, 2022). Carbon finance is a new trend in global carbon markets, with many legal and policy complexities. Applying EU rules on carbon finance to the specialized regulatory regime of China's carbon market can help it function (Chen and Wu, 2023). More importantly, Chinese firms' total factor productivity (TFP) would increase approximately 22.73 percent if China's carbon emissions price equals that of the EU ETS (Wu and Wang, 2022). The sensitivity of the Chinese market to cross-border carbon policy, especially in light of the European Union Carbon Border Adjustment Mechanism (EU CBAM), is a matter of concern, however, as China's exports to the EU are carbon-intensive. Carbon policy adjustment and reconciliation may thus be required to avoid affecting EU-China trade (Shen *et al.*, 2023).

The carbon trading market is valuable for attracting investors' attention in CCUS development initiatives (Lin and Tan, 2021), which are critical to tackling climate change (McLaughlin *et al.*, 2023; Roy *et al.*, 2023). As mentioned above, however, CCUS is still in the demonstration stage, and there are many uncertainties in business model and policy incentives that traditional approaches can no longer handle (Ye *et al.*, 2022). Efforts such as regional corridors for optimization of scale, societal and stakeholder engagement, carbon pricing, and public–private partnerships for financing, knowledge, and technology transfer, are crucial for effective CCSU installations (Lau *et al.*, 2021).

3. Status of CCUS in China

China is the first country to promote using CCUS to reduce CO_2 emissions globally (Qin *et al.*, 2015), and has the most comprehensive experience of CCUS to balance economic expansion and carbon emissions (Yao *et al.*, 2023). CCUS also helps China avoid stranded assets and draws business community support against climate change. China's carbon storage capacity is estimated at 3120 $GtCO_2$, indicating CCUS potential (Xu and Dai, 2021). Chen *et al.* (2022) show that the CCUS "Golden Age" will be 2040–2060 globally, versus 2030–2050 in China. As mentioned before, however, CCUS still lags in China, stressing the importance of continuous government support as a component of energy efficiency and low-carbon transition plans (Jiang and Ashworth, 2021).

CCUS has in fact been a vital part of Chinese government policy over the past two decades. In 2007, the government enlisted CCUS for carbon reduction, and commenced demonstration projects in 2013, including environmental impact assessment and monitoring. In 2017, the government concentrated on CCUS technology development, which was boosted in 2021 by the aforementioned CCUS pilots (Xu and Dai, 2021). These initiatives led to expectations for CCUS at various economic and political levels (Yin and Xu, 2022). Ma *et al.* (2023) report areas needing improvement in China's multi-department generated CCUS policy. First, they show that supply-side policy needs more balance, as it currently depends on technology research and development. Second, environmental policies must be made more relevant and applicable, especially legislation, standards, and incentives. Finally, policies that drive CCUS demand in national and international markets alike are also required.

Maximal CCUS development can help China achieve carbon neutrality by 2054–2058, as opposed to 2061–2064 for lesser degrees of development (Sun *et al.*, 2022). Chinese oil companies have been using CO_2 storage technologies for many years, constantly producing positive results, despite a lack of large-scale CCUS adoptions (Che *et al.*, 2022). Wei *et al.* (2022) show that CCUS will also benefit China's steel industry on the order of 17.6 percent–56.5 percent, contingent upon carbon price mechanisms or supportive government policies, making it catalyze a green transition for the sector. Han *et al.* (2023) further point out that the three-dimensional model of the Chinese CCUS coal-based power plant can reach a fitness level of 0.99 with a benefit ratio of 5:9:1, corresponding to the economy, the environment, and energy, respectively. CCUS can similarly benefit China's chemical and cement industries (Guo *et al.*, 2023; Xie *et al.*, 2022).They project that a CCUS pipeline network model would potentially remove 280 $MtCO_2$ annually from the atmosphere, with net income of USD 7.85 per ton of CO_2, or USD 2.20 billion annual total (Xie *et al.*, 2022).

The public and stakeholders influence strategic CCUS investment decisions (Liu *et al.*, 2021; Sitinjak *et al.*, 2023), as these are sensitive technologies that require public understanding and acceptance. Even if a significant percentage of mainstream Chinese news reports present CCUS positively, more is needed to address misperceptions and depict the subject comprehensively (Jiang *et al.*, 2022). Heightened awareness promotes mass acceptance, which in China is also more strongly affected by perceived benefits than perceived risks. Chinese authorities and leading clean energy players should therefore make greater efforts in CCUS-related publicity and incentives if they want to make it effective at reducing carbon emissions (Liu *et al.*, 2021).

4. Methodology

The jump approach has received attention in recent studies (Dutta, 2017; Dutta *et al.*, 2022; Dutta and Das, 2022; Fowowe, 2013; Gronwald, 2019; Kuttu, 2017; Zhou *et al.*, 2019). Following Chan and Maheu (2002), we accordingly employ the GARCH-jump model, as follows[2]:

$$r_t = \pi + \mu r_{t-1} + \varepsilon_t \tag{1}$$

---

[2]The AR(1) specification is selected based on the Akaike criterion.

where r_t indicates the logarithmic difference for the CEP index at time t, and ε_t refers to the innovation term, specified as:

$$\varepsilon_t = \varepsilon_{1t} + \varepsilon_{2t} \qquad (2)$$

where ε_{1t} will follow the GARCH (1,1) specification:

$$\varepsilon_{1t} = \sqrt{h_t}\, z_t, \quad z_t \sim NID(0,1) \qquad (3)$$
$$h_t = \omega + \alpha \varepsilon_{1t-1}^2 + \beta h_{t-1}$$

and ε_{2t} denotes a jump innovation, defined as:

$$\varepsilon_{2t} = \sum_{l=1}^{n_t} J_{tl} - \theta \lambda_t \qquad (4)$$

where J_{tl} is the jump size with a mean value θ and a variance ϑ^2, $\sum_{l=1}^{n_t} J_{tl}$ refers to the jump factor, and n_t represents the jump frequency at time t, following a Poisson distribution given by:

$$P(n_t = j|I_{t-1}) = \frac{e^{-\lambda_t} \lambda_t^j}{j!}, \quad j = 0,1,2,\ldots \qquad (5)$$

with an autoregressive conditional jump intensity (ARJI), given as:

$$\lambda_t = \lambda_0 + \rho \lambda_{t-1} + \gamma \xi_{t-1} \qquad (6)$$

In Equation (6), λ_t indicates the time-varying conditional jump intensity parameter, λ_0 is the constant jump intensity, and ξ_{t-1} denotes the intensity residual. Chan and Maheu (2002) assume that $\lambda_t > 0$, $\lambda_0 > 0$, $\rho > 0$, and $\gamma > 0$.

We define the log-likelihood as:

$$L(\Theta) = \sum_{t=1}^{T} \log f(X_t|I_{t-1};\Theta)$$

where $\Theta = (\pi, \mu, \omega, \alpha, \beta, \theta, \vartheta, \lambda_0, \rho, \gamma)$ and I_{t-1} is the information set.

Note that the jump-induced volatility (JV) is given as:

$$JV_t = \sqrt{(\theta^2 + \vartheta^2)\lambda_t} \qquad (7)$$

Next, we examine whether the OVX, clean energy asset price volatility, and EU emissions price volatility explain JV, as defined in Equation (7). To serve this purpose, we estimate the following model:

$$JV_t = \varphi_0 + \varphi_1 U_{t-1} + \varepsilon_t \tag{8}$$

where U indicates a vector of uncertainty indexes. If φ_1 deviates significantly from zero for a particular uncertainty measure, we can infer that it may explain the aforementioned jump-induced risk.[3]

5. Data

In this study, we use information on the Shenzhen carbon emission trading price (CEP) for its quality and duration. While there are currently eight carbon emission trading pilot programs in China, the Shenzhen carbon emission trading market has received attention in recent studies (Chen and Xu, 2022; Zheng *et al.*, 2021). Our sample includes daily observations spanning from August 2013 to August 2022. As mentioned earlier, our analysis has considered several influences, including OVX, EPU, and GPR. The OVX data are collected from the DataStream database, and the GPR index is found on the EPU website, to which an index for China has been recently added by Lee *et al.* (2023). We obtained the data from their website, twitterchnepu.github.io.

Table 1 shows descriptive statistics for emission prices, OVX, EPU, and GPR, indicating that carbon market returns skew positive, which is also true for all uncertainty indicators. All series are found to be leptokurtic and hence violate the normality assumption. The Jarque–Bera test gives the same result.

Table 2 displays the outcomes of the augmented Dickey–Fuller (ADF) and Phillips–Perron (PP) unit root tests. The null hypothesis of these tests is that the data follow a non-stationary process, with results reported for both levels and first differences. These findings suggest that all uncertainty measures are stationary even at levels, while ADF shows the carbon price index appears non-stationary at levels. Both ADF and PP find the CEP index stationary after adjusting for logarithmic differences.

Figure 1 depicts the Chinese emissions price index, demonstrating that carbon prices fell during the COVID-19 pandemic and rose following

---

[3] Many recent studies have found significant linkages between the Chinese carbon market and global energy markets; see, for example, Ji *et al.* (2019), Lin and Chen (2019), Zhu *et al.* (2020), Gong *et al.* (2021), and Xu (2021).

Table 1.   Descriptive statistics.

	CEP returns	OVX	EPU	GPR
Mean	0.0033	36.6066	135.2457	145.4613
Std. Dev.	0.3389	13.61	267.55	1152.40
Skewness	0.3915	2.23	6.66	35.74
Kurtosis	23.03	20.65	71.72	1346.66
Jarque-Bera	30942.00***	24532.16***	362653.30***	134000.01***

Note: This table presents summary statistics for all variables. *** indicates statistical significance at $p = 0.01$.

Table 2.   Unit root tests.

	Augmented Dickey–Fuller test		Phillips–Perron test	
	Level	1st Difference	Level	1st Difference
CEP	−2.55	−27.59***	−4.51***	−83.93***
OVX	−4.40***	−23.34***	−6.15***	−59.21***
EPU	−8.45***	−21.82***	−40.30***	−394.70***
GPR	−42.16***	−19.34***	−42.15***	1781.14***

Note: This table presents unit root test results for all variables. We consider only logarithmic differences in CEP. *** indicates statistical significance at $p = 0.01$.

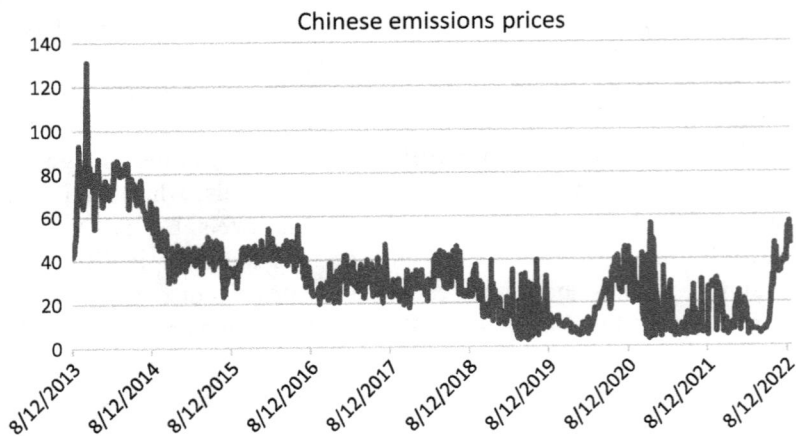

Figure 1.   Chinese emissions price index.

Russia's invasion of Ukraine. Carbon prices also dropped during the 2014–2015 downturn in global crude oil markets. Chinese carbon prices correlate closely with traditional energy prices.

6. Results

6.1. *GARCH-jump process findings*

Table 3 shows the GARCH-jump process estimates, along with the GARCH(1,1) model and constant jump intensity (CJI) process estimates for comparison. The CJI process, developed by Jorion (1988), assumes that $\lambda_t = \lambda_0$, i.e., the jump intensity parameter is constant over time.

The results show that the GARCH parameters (α, β) are statistically significant across all models. We also notice that the sum of $\hat{\alpha}$ and $\hat{\beta}$ are close to unity, indicating persistent volatility. Turning to the CJI process, we find that the mean parameter (θ) for jumps is insignificant, whereas the variance parameter (ϑ) is statistically significant. The positive coefficient

Table 3.   GARCH model estimates.

Models →	GARCH (1,1)		CJI		ARJI	
	Estimate	S.E.	Estimate	S.E.	Estimate	S.E.
π	0.0008	0.0020	−0.0015	0.0016	−0.0023	0.0016
μ	−0.3252***	0.0255	−0.2606***	0.0231	−0.2407***	0.0243
ω	0.0001***	0.00002	0.0003***	0.0001	0.0007***	0.0001
α	0.1229***	0.0108	0.3276***	0.0347	0.4021***	0.0415
β	0.8952***	0.0072	0.6365***	0.0297	0.4831***	0.0430
θ			0.0796	0.0886	0.0089	0.0538
ϑ^2			0.7190***	0.0975	0.6518***	0.0725
λ_0			0.0392***	0.0081	0.0002	0.0002
ρ					0.9973***	0.0025
γ					0.0456***	0.0147
Log-likelihood	853.07		1034.36		1070.65	

Note: This table presents Chinese emissions index GARCH model findings. *** indicates statistical significance at $p = 0.01$.

of the jump variance indicates that carbon price volatility increases with increments in volatility caused by anomalous information.

Turning to the GARCH-ARJI process, we observe that the parameter λ_0, which appears to be significant in the CJI model, is insignificant here, and that the ARJI parameters (ρ, γ) are strongly significant at $p = 0.01$, suggesting that jumps exist in the Chinese carbon market and that such jumps evolve over time. Note that the parameter $\rho = 0.9973$ reveals that jump intensity is highly persistent. The significant value for γ also suggests a diminishing effect of 0.0456 on future jump intensity. The ARJI parameters (ρ, γ) also exceed zero, satisfying the positivity constraints. The log-likelihood values shown in the last row also confirm the superiority of this process.

Figure 2 depicts Chinese carbon market jump intensities, showing that they tend to be time dependent. More importantly, λ_t seems to increase in times of crisis, such throughout the COVID-19 pandemic. Previous studies (Dutta *et al.*, 2021; Dutta *et al.*, 2022) also claim that jumps in energy prices tend to rise amid market downturns.

Our findings are consistent with earlier research. Dutta (2018), for instance, shows that time-varying jumps represent a common event in the EU carbon emissions market, and that modeling such jumps plays a pivotal role in more precisely understanding the carbon price volatility. We therefore recommend the GARCH-ARJI process when studying carbon price volatility dynamics in China.

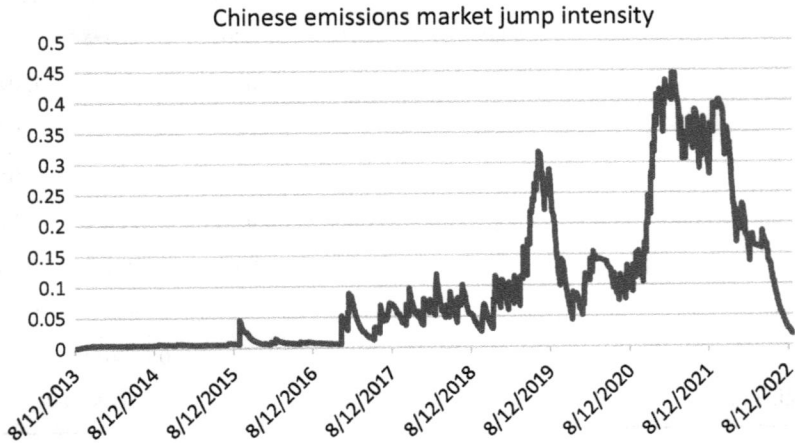

Figure 2.   Chinese emissions price index jump intensity.

6.2. *JV determinants*

In this section, we examine whether leading uncertainty measures may explain JV in the Chinese emissions market. Table 4 shows the impacts of OVX, EPU, and GPR on JV, all of which are substantial, as the corresponding parameters are statistically significant at $p = 0.01$. Notably, OVX and EPU influence JV positively, whereas GPR has a negative impact. The R^2 statistics show that JV is better explained in China by EPU than OVX or GPR. These linkages could be due to the high correlations between energy and emission prices, geopolitics, oil, and inflation, investors' concern for climate change, and financial market integration.

Our findings are in line with earlier literature (Dutta, 2018; Viteva *et al.*, 2014). Dutta (2018) also shows that OVX exerts a positive impact on the volatility of EU emission prices. One plausible explanation is that rising OVX tends to drive down energy prices, which in turn promotes use of traditional energy sources, resulting in increased emission price volatility. This finding also establishes that energy and emission prices move in the same direction. We also observe that EU and Chinese emission prices react to OVX in a similar manner. The positive association between EPU and JV implies that rising EPU tends to raise policy uncertainty for sustainable investments, which raise volatility levels in the Chinese the emissions market, while the negative association between GPR and JV could be attributed to the fact that as GPR rises, crude oil, which is highly

Table 4.   JV determinants.

	Estimate	Standard Error	*t*-statistic	*p*-value	R^2
Panel A: OVX					39.54
Constant	0.0302***	0.0049	6.08	0.00	
φ_1	0.0008***	0.0001	8.21	0.00	
Panel B: EPU					41.49
Constant	0.0495***	0.0019	25.99	0.00	
φ_1	0.0001***	0.00001	10.01	0.00	
Panel C: GPR					38.61
Constant	0.1093***	0.0178	6.12	0.00	
φ_1	−0.0111***	0.0038	−2.89	0.00	

Notes: φ_1 measures OVX/EPU/GPR impact on emissions market JV. ***indicates statistical significance at p –0.01.

sensitive to such risk (Tiwari *et al.*, 2020), seems to be replaced by renewable energies, leading to a rise in emissions prices and a significant drop in volatility.

7. Additional Tests: Do Jumps Exist for Outlier-Free Carbon Prices?

We now explore whether Chinese carbon prices involve outliers, and if jumps still exist after correcting for these. We employ the methodology proposed by Ané *et al.* (2008) to identify them.

Suppose, R_t denotes the log return by carbon price on day t. Then the AR(1)-GARCH(1,1) process is defined as:

$$R_t = b_0 + b_1 R_{t-1} + \varepsilon_t \tag{9}$$

$$\sigma_t^2 = a_0 + a_1 \varepsilon_{t-1}^2 + a_2 \sigma_{t-1}^2 \tag{10}$$

where $\varepsilon_t = \sigma_t z_t$ with z_t being an i.i.d. process such as $z_t / I_{t-1} \sim IIN(0,1)$; I_{t-1} indicates the filtration of information at time $t-1$.

Then R_{t+1} is assumed to be an outlier if it does not fall in the following interval:

$$R_{t+1} \in \left[R_{t,t+1} \pm F \left(1 - \frac{\alpha}{2} \right) \sigma_{t,t+1} \right]$$

where, $R_{t,t+1}$ refers to the one-step ahead prediction, which is modeled as:

$$R_{t,t+1} = E(R_{t+1}/I_t) = b_0 + b_1 R_t + b_2 R_{t-1}$$

and $\sigma_{t,t+1}^2$ is the one-step ahead risk prediction, specified as:

$$\sigma_{t,t+1}^2 = \operatorname{var}(R_{t+1}/I_t) = a_0 + (a_1 + a_2)\sigma_t^2$$

Additionally, $F\left(1 - \frac{\alpha}{2}\right) = P(z_t \leq 1 - \%)$ refers to a fractile of the presumed conditional distribution. This process keeps rolling over until the end of our sample.

Our analysis detects several such outliers, which mainly occur during such aforementioned crises as the 2014 oil market downturn and COVID-19 pandemic. We now re-estimate the jump process after deleting these outliers from our original dataset and report the results in Table 5, finding

Table 5. Testing for jumps on outlier-free carbon market data.

Models →	CJI		ARJI	
	Estimate	S.E.	Estimate	S.E.
π	−0.0020	0.0016	−0.0026***	0.0004
μ	−0.2819***	0.0234	−0.2573***	0.0271
ω	0.0003***	0.0001	0.0008***	0.0002
α	0.3224***	0.0346	0.4002***	0.0418
β	0.6468***	0.0287	0.4758***	0.0392
θ	0.1034	0.1012	0.0134	0.0482
ϑ^2	0.7137***	0.0872	0.6562***	0.0414
λ_0	0.0395***	0.0083	0.0004***	0.0001
ρ			1.0002***	0.0004
γ			0.0062***	0.0002
Log-likelihood	955.76		1000.32	

Note: This table presents Chinese emissions index jump model findings after correcting for outliers. *** indicates statistical significance at $p = 0.01$.

that time-dependent jumps exist even after removing the outliers. These results hold for both CJI and ARJI. Hence, we conclude that modeling jumps is important for risk assessment, and detecting such jumps is also crucial for efficient carbon trading in China.

8. Conclusions and Policy Implications

While CCUS is now considered one of the most efficient means for achieving net-zero targets, rising economic uncertainty, geopolitical conflict, and oil price volatility slow its deployment. While carbon emissions trading mechanisms can effectively promote CCUS, recent literature demonstrates that emissions trading schemes are still immature in developing economies such as China, where carbon pricing appears a key strategy to reduce CO_2 energy emissions. In this study, we thus investigate the Chinese carbon market's volatility dynamics, with particular attention to time-varying jumps in emissions prices. The reason is that, as JV represents an important element of risk, precise information on such risk is essential for more efficient carbon trading. Employing the GARCH-jump process, we find that time-dependent jumps do occur in the Chinese emissions market and that key uncertainty indicators, including economic policy uncertainty, OVX, and GPR, can successfully explain JV in this market.

Our study has significant implications for investors, given that precise financial market volatility estimates play a key role in risk management. It suggests that, when investing in the Chinese emissions market, financial institutions should pay careful attention to time-varying jumps in carbon prices as JV may carry important information about potential risk. As we find that key uncertainty indicators may explain JV, information about these may also be useful for market participants making investment decisions when JV increases, making our analysis useful in deriving appropriate asset pricing models that may minimize carbon price uncertainty.

This study also has implications for policymakers. First, the Chinese government should focus on greater carbon market efficiency, as this is crucial for improved CCUS deployment in China. Identifying time-dependent jumps in emission prices will help policymakers in this regard, which will also help attract eco-friendly investors, another key to CO_2 emissions reductions. Second, oil market stability, which is another prerequisite to success with CCUS, will require such actions as adding to oil reserves, updating crude oil futures monitoring policy, and raising carbon taxes. Renewable energy will also help by reducing the impact of oil price uncertainty and geopolitical conflict. Third, the government strategy for encouraging sustainable investments must also concentrate on other promotional efforts, including financial incentives and higher carbon taxes, in addition to CCUS deployment. Finally, the Chinese government should also increase its CCUS R&D budget.

References

Ané, T., Ureche-Rangau, L., Gambet, J.-B., and Bouverot, J. (2008). Robust outlier detection for Asia–Pacific stock index returns. *Journal of International Financial Markets, Institutions and Money*, *18*(4), 326–343. https://doi.org/10.1016/j.intfin.2007.03.001.

Chan, W. H. and Maheu, J. M. (2002). Conditional jump dynamics in stock market returns. *Journal of Business & Economic Statistics*, *20*(3), 377–389. https://doi.org/10.1198/073500102288618513.

Che, X., Yi, X., Dai, Z., Zhang, Z., and Zhang, Y. (2022). Application and development countermeasures of CCUS technology in China's petroleum industry. *Atmosphere*, *13*(11), 1757. https://doi.org/10.3390/atmos13111757.

Chen, B. and Wu, R. (2023). Legal and policy pathways of carbon finance: Comparative analysis of the carbon market in the EU and China. *European Business Organization Law Review, 24*(1), 377–389. https://doi.org/10.1007/s40804-022-00259-x.

Chen, H. and Xu, C. (2022). The impact of cryptocurrencies on China's carbon price variation during COVID-19: A quantile perspective. *Technological Forecasting and Social Change, 183,* 121933. https://doi.org/10.1016/j.techfore.2022.121933.

Chen, S., Liu, J., Zhang, Q., Teng, F., and McLellan, B. C. (2022). A critical review on deployment planning and risk analysis of carbon capture, utilization, and storage (CCUS) toward carbon neutrality. *Renewable and Sustainable Energy Reviews, 167,* 112537. https://doi.org/10.1016/j.rser.2022.112537.

Chun, S. (2018). Spillover effects of price fluctuation on China's carbon market and EU carbon market. *The Journal of Industrial Economics,* 37, 97–105.

Compernolle, T., Kort, P. M., and Thijssen, J. J. J. (2022). The effectiveness of carbon pricing: The role of diversification in a firm's investment decision. *Energy Economics, 112,* 106115. https://doi.org/10.1016/j.eneco.2022.106115.

Dutta, A. (2017). Oil price uncertainty and clean energy stock returns: New evidence from crude oil volatility index. *Journal of Cleaner Production, 164,* 1157–1166. https://doi.org/10.1016/j.jclepro.2017.07.050.

Dutta, A. (2018). Modeling and forecasting the volatility of carbon emission market: The role of outliers, time-varying jumps and oil price risk. *Journal of Cleaner Production, 172,* 2772–2781. https://doi.org/10.1016/j.jclepro.2017.11.135.

Dutta, A., Bouri, E., Saeed, T., and Vo, X. V. (2021). Crude oil volatility and the biodiesel feedstock market in Malaysia during the 2014 oil price decline and the COVID-19 outbreak. *Fuel, 292,* 120221. https://doi.org/10.1016/j.fuel.2021.120221.

Dutta, A. and Das, D. (2022). Forecasting realized volatility: New evidence from time-varying jumps in VIX. *Journal of Futures Markets, 42*(12), 2165–2189. https://doi.org/10.1002/fut.22372.

Dutta, A., Soytas, U., Das, D., and Bhattacharyya, A. (2022). In search of time-varying jumps during the turmoil periods: Evidence from crude oil futures markets. *Energy Economics, 114,* 106275. https://doi.org/10.1016/j.eneco.2022.106275.

Fang, G., Tian, L., Liu, M., Fu, M., and Sun, M. (2018). How to optimize the development of carbon trading in China—Enlightenment from evolution rules of the EU carbon price. *Applied Energy, 211,* 1039–1049. https://doi.org/10.1016/j.apenergy.2017.12.001.

Fowowe, B. (2013). Jump dynamics in the relationship between oil prices and the stock market: Evidence from Nigeria. *Energy, 56,* 31–38. https://doi.org/10.1016/j.energy.2013.04.062.

Gronwald, M. (2019). Is Bitcoin a Commodity? On price jumps, demand shocks, and certainty of supply. *Journal of International Money and Finance, 97*, 86–92. https://doi.org/10.1016/j.jimonfin.2019.06.006.

Guo, Y., Luo, L., Liu, T., Hao, L., Li, Y., Liu, P., and Zhu, T. (2023). A review of low-carbon technologies and projects for the global cement industry. *Journal of Environmental Sciences, 136*, 682–697. https://doi.org/10.1016/j.jes.2023.01.021.

Han, J., Li, J., Tang, X., Wang, L., Yang, X., Ge, Z., and Yuan, F. (2023). Coal-fired power plant CCUS project comprehensive benefit evaluation and forecasting model study. *Journal of Cleaner Production, 385*, 135657. https://doi.org/10.1016/j.jclepro.2022.135657.

Hua, Y. and Dong, F. (2019). China's carbon market development and carbon market connection: A literature review. *Energies, 12*(9), 1663. https://doi.org/10.3390/en12091663.

Ji, Q., Xia, T., Liu, F., and Xu, J.-H. (2019). The information spillover between carbon price and power sector returns: Evidence from the major European electricity companies. *Journal of Cleaner Production, 208*, 1178–1187. https://doi.org/10.1016/j.jclepro.2018.10.167.

Jiang, K. and Ashworth, P. (2021). The development of Carbon Capture Utilization and Storage (CCUS) research in China: A bibliometric perspective. *Renewable and Sustainable Energy Reviews, 138*, 110521. https://doi.org/10.1016/j.rser.2020.110521.

Jiang, K., Ashworth, P., Zhang, S., and Hu, G. (2022). Print media representations of carbon capture utilization and storage (CCUS) technology in China. *Renewable and Sustainable Energy Reviews, 155*, 111938. https://doi.org/10.1016/j.rser.2021.111938.

Jorion, P. (1988). On jump processes in the foreign exchange and stock markets. *Review of Financial Studies, 1*(4), 427–445. https://doi.org/10.1093/rfs/1.4.427.

Känzig, D. (2023). *The Unequal Economic Consequences of Carbon Pricing* (No. 31221).

Khan, J. and Johansson, B. (2022). Adoption, implementation and design of carbon pricing policy instruments. *Energy Strategy Reviews, 40*, 100801. https://doi.org/10.1016/j.esr.2022.100801.

Ko, I. and Lee, T. (2022). Carbon pricing and decoupling between greenhouse gas emissions and economic growth: A panel study of 29 European countries, 1996–2014. *Review of Policy Research, 39*(5), 651–673. https://doi.org/10.1111/ropr.12458.

Kojima, S. and Asakawa, K. (2021). *Expectations for Carbon Pricing in Japan in the Global Climate Policy Context* (T. Arimura & S. Matsumoto, Eds.), 1–21. Springer. https://doi.org/10.1007/978-981-15-6964-7_1.

Kuttu, S. (2017). Time-varying conditional discrete jumps in emerging African equity markets. *Global Finance Journal, 32*, 35–54. https://doi.org/10.1016/j. gfj.2016.06.004.

Landis, F., Fredriksson, G., and Rausch, S. (2021). Between- and within-country distributional impacts from harmonizing carbon prices in the EU. *Energy Economics, 103*, 105585. https://doi.org/10.1016/j.eneco.2021.105585.

Lau, H. C., Ramakrishna, S., Zhang, K., and Radhamani, A. V. (2021). The role of carbon capture and storage in the energy transition. *Energy & Fuels, 35*(9), 7364–7386. https://doi.org/10.1021/acs.energyfuels.1c00032.

Lee, K., Choi, E., and Kim, M. (2023). Twitter-based Chinese economic policy uncertainty. *Finance Research Letters, 53*, 103627. https://doi.org/10.1016/j. frl.2023.103627.

Li, M., Zhang, D., Li, C.-T., Mulvaney, K. M., Selin, N. E., and Karplus, V. J. (2018). Air quality co-benefits of carbon pricing in China. *Nature Climate Change, 8*(5), 398–403. https://doi.org/10.1038/s41558-018-0139-4.

Liao, H., Wu, D., Wang, Y., Lyu, Z., Sun, H., Nie, Y., and He, H. (2022). Impacts of carbon trading mechanism on closed-loop supply chain: A case study of stringer pallet remanufacturing. *Socio-Economic Planning Sciences, 81*, 101209. https://doi.org/10.1016/j.seps.2021.101209.

Lin, B. and Chen, Y. (2019). Dynamic linkages and spillover effects between CET market, coal market and stock market of new energy companies: A case of Beijing CET market in China. *Energy, 172*, 1198–1210. https://doi. org/10.1016/j.energy.2019.02.029.

Lin, B. and Jia, Z. (2019). Impacts of carbon price level in carbon emission trading market. *Applied Energy, 239*, 157–170. https://doi.org/10.1016/j. apenergy.2019.01.194.

Lin, B. and Tan, Z. (2021). How much impact will low oil price and carbon trading mechanism have on the value of carbon capture utilization and storage (CCUS) project? Analysis based on real option method. *Journal of Cleaner Production, 298*, 126768. https://doi.org/10.1016/j.jclepro.2021.126768.

Liu, B., Xu, Y., Yang, Y., and Lu, S. (2021). How public cognition influences public acceptance of CCUS in China: Based on the ABC (affect, behavior, and cognition) model of attitudes. *Energy Policy, 156*, 112390. https://doi. org/10.1016/j.enpol.2021.112390.

Liu, J., Hou, J., Fan, Q., and Chen, H. (2022). China's national ETS: Global and local lessons. *Energy Reports, 8*, 428–437. https://doi.org/10.1016/j. egyr.2022.03.097.

Ma, Q., Wang, S., Fu, Y., Zhou, W., Shi, M., Peng, X., Lv, H., Zhao, W., and Zhang, X. (2023). China's policy framework for carbon capture, utilization and storage: Review, analysis, and outlook. *Frontiers in Energy, 17*, 400–411. https://doi.org/10.1007/s11708-023-0862-z.

McLaughlin, H., Littlefield, A. A., Menefee, M., Kinzer, A., Hull, T., Sovacool, B. K., Bazilian, M. D., Kim, J., and Griffiths, S. (2023). Carbon capture utilization and storage in review: Sociotechnical implications for a carbon reliant world. *Renewable and Sustainable Energy Reviews, 177*, 113215. https://doi.org/10.1016/j.rser.2023.113215.

Mo, J., Tu, Q., and Wang, J. (2023). Carbon pricing and enterprise productivity-The role of price stabilization mechanism. *Energy Economics, 120*, 106631. https://doi.org/10.1016/j.eneco.2023.106631.

Narassimhan, E., Gallagher, K. S., Koester, S., and Alejo, J. R. (2018). Carbon pricing in practice: A review of existing emissions trading systems. *Climate Policy, 18*(8), 967–991. https://doi.org/10.1080/14693062.2018.1467827.

Qi, S., Cheng, S., Tan, X., Feng, S., and Zhou, Q. (2022). Predicting China's carbon price based on a multi-scale integrated model. *Applied Energy, 324*, 119784. https://doi.org/10.1016/j.apenergy.2022.119784.

Roy, P., Mohanty, A. K., and Misra, M. (2023). Prospects of carbon capture, utilization and storage for mitigating climate change. *Environmental Science: Advances, 2*(3), 409–423. https://doi.org/10.1039/D2VA00236A

Shen, H., Yang, Q., Luo, L., and Huang, N. (2023). Market reactions to a cross-border carbon policy: Evidence from listed Chinese companies. *The British Accounting Review, 55*(1), 101116. https://doi.org/10.1016/j.bar.2022.101116.

Sitinjak, C., Ebennezer, S., and Ober, J. (2023). Exploring public attitudes and acceptance of CCUS technologies in JABODETABEK: A cross-sectional study. *Energies, 16*(10), 4026. https://doi.org/10.3390/en16104026.

Sun, L.-L., Cui, H.-J., and Ge, Q.-S. (2022). Will China achieve its 2060 carbon neutral commitment from the provincial perspective? *Advances in Climate Change Research, 13*(2), 169–178. https://doi.org/10.1016/j.accre.2022.02.002.

Tiwari, A. K., Aye, G. C., Gupta, R., and Gkillas, K. (2020). Gold-oil dependence dynamics and the role of geopolitical risks: Evidence from a Markov-switching time-varying copula model. *Energy Economics, 88*, 104748. https://doi.org/10.1016/j.eneco.2020.104748.

Thube, S., Peterson, S., Nachtigall, D., and Ellis, J. (2021). The economic and environment benefits from international co-ordination on carbon pricing: a review of economic modelling studies. *Environmental Research Letters, 16*(11), 113002. https://doi.org/10.1088/1748-9326/ac2b61.

van den Bergh, J., and Savin, I. (2021). Impact of carbon pricing on low-carbon innovation and deep decarbonisation: Controversies and path forward. *Environmental and Resource Economics, 80*(4), 705–715. https://doi.org/10.1007/s10640-021-00594-6.

Venmans, F., Ellis, J., and Nachtigall, D. (2020). Carbon pricing and competitiveness: Are they at odds? *Climate Policy, 20*(9), 1070–1091. https://doi.org/10.1080/14693062.2020.1805291.

Viteva, S., Veld-Merkoulova, Y. V., and Campbell, K. (2014). The forecasting accuracy of implied volatility from ECX carbon options. *Energy Economics, 45*, 475–484. https://doi.org/10.1016/j.eneco.2014.08.005.

Wang, Y. and Guo, Z. (2018). The dynamic spillover between carbon and energy markets: New evidence. *Energy, 149*, 24–33. https://doi.org/10.1016/j.energy.2018.01.145.

Wei, X., Mohsin, M., and Zhang, Q. (2022). Role of foreign direct investment and economic growth in renewable energy development. *Renewable Energy, 192*, 828–837. https://doi.org/10.1016/j.renene.2022.04.062.

Wen, F., Zhao, H., Zhao, L., and Yin, H. (2022). What drive carbon price dynamics in China? *International Review of Financial Analysis, 79*, 101999. https://doi.org/10.1016/j.irfa.2021.101999.

Weng, Q. and Xu, H. (2018). A review of China's carbon trading market. *Renewable and Sustainable Energy Reviews, 91*, 613–619. https://doi.org/10.1016/j.rser.2018.04.026.

Wu, Q. and Wang, Y. (2022). How does carbon emission price stimulate enterprises' total factor productivity? Insights from China's emission trading scheme pilots. *Energy Economics, 109*, 105990. https://doi.org/10.1016/j.eneco.2022.105990.

Xie, J., Li, X., and Gao, X. (2022). Pipeline network options of CCUS in coal chemical industry. *Atmosphere, 13*(11), 1864. https://doi.org/10.3390/atmos13111864.

Xu, S. and Dai, S. (2021). CCUS As a second-best choice for China's carbon neutrality: An institutional analysis. *Climate Policy, 21*(7), 927–938. https://doi.org/10.1080/14693062.2021.1947766.

Xu, Y. (2021). Risk spillover from energy market uncertainties to the Chinese carbon market. *Pacific-Basin Finance Journal, 67*, 101561. https://doi.org/10.1016/j.pacfin.2021.101561.

Yan, J. and Yang, J. (2021). Carbon pricing and income inequality: A case study of Guangdong Province, China. *Journal of Cleaner Production, 296*, 126491. https://doi.org/10.1016/j.jclepro.2021.126491.

Yao, J., Han, H., Yang, Y., Song, Y., and Li, G. (2023). A review of recent progress of carbon capture, utilization, and storage (CCUS) in China. *Applied Sciences, 13*(2), 1169. https://doi.org/10.3390/app13021169.

Ye, J., Yan, L., Liu, X., and Wei, F. (2022). Economic feasibility and policy incentive analysis of carbon capture, utilization, and storage (CCUS) in coal-fired power plants based on system dynamics. *Environmental Science and Pollution Research, 30*(13), 37487–37515. https://doi.org/10.1007/s11356-022-24888-4.

Yin, C. and Xu, H. (2022). Assessing the niche development of carbon capture and storage through strategic niche management approach: The case of China. *International Journal of Greenhouse Gas Control, 119*, 103721. https://doi.org/10.1016/j.ijggc.2022.103721.

Zeng, Y., Weishaar, S. E., and Couwenberg, O. (2016). Absolute vs. intensity-based caps for carbon emissions target setting. *European Journal of Risk Regulation*, *7*(4), 764–781. https://doi.org/10.1017/S1867299X00010187.

Zhao, Y., Wang, C., and Cai, W. (2022). Carbon pricing policy, revenue recycling schemes, and income inequality: A multi-regional dynamic CGE assessment for China. *Resources, Conservation and Recycling*, *181*, 106246. https://doi.org/10.1016/j.resconrec.2022.106246.

Zheng, Y., Zhou, M., and Wen, F. (2021). Asymmetric effects of oil shocks on carbon allowance price: Evidence from China. *Energy Economics*, *97*, 105183. https://doi.org/10.1016/j.eneco.2021.105183.

Zhou, C., Wu, C., & Wang, Y. (2019). Dynamic portfolio allocation with time-varying jump risk. *Journal of Empirical Finance*, *50*, 113–124. https://doi.org/10.1016/j.jempfin.2019.01.003.

Zhu, B., Huang, L., Yuan, L., Ye, S., and Wang, P. (2020). Exploring the risk spillover effects between carbon market and electricity market: A bidimensional empirical mode decomposition based conditional value at risk approach. *International Review of Economics & Finance*, *67*, 163–175. https://doi.org/10.1016/j.iref.2020.01.003.

Chapter 9

The Impact of CCUS Innovation on Green Total Factor Productivity in China

Kangyin Dong[*,§]**, Jianda Wang**[*,¶]**,**
Congyu Zhao[*,‖]**, Farhad Taghizadeh-Hesary**[†,**]**,**
and Han Phoumin[‡,††]

School of International Trade and Economics,
University of International Business
and Economics, Beijing, China

†*School of Global Studies and Tokai Research*
Institute for Environment and Sustainability (TRIES),
Tokai University, Tokyo, Japan

‡*Economic Research Institute for ASEAN*
and East Asia (ERIA), Jakarta, Indonesia

§*dongkangyin@uibe.edu.cn*

¶*wangjd1993@163.com*

‖*cyzhao1998@163.com*

**farhad@tsc.u-tokai.ac.jp*

††*han.phoumin@eria.org*

Abstract

Technology innovation encompasses the core competencies of research and development (R&D). The significance of carbon capture, utilization, and storage (CCUS) innovation during low-carbon transition is self-evident. We investigate the impact of CCUS technology innovation on green total factor productivity (GTFP) based on the 2007–2019 panel dataset. We also examine their nonlinear nexus and identify potential impact channels. Our findings show that CCUS innovation significantly improves GTFP by promoting industrial infrastructure and carbon emissions efficiency, environmental regulation positively moderates the nexus between CCUS technology innovation and GTFP, and CCUS technology innovation is more effective in promoting GTFP in regions that already have high GTFP. We conclude with specific policy implications for better CCUS technology innovation and GTFP development based on these findings.

Keywords: CCUS technology innovation, green total factor productivity, moderating analysis, mediating effect, China.

1. Introduction

The issue of climate change has garnered significant attention from governments and scholars worldwide, with a primary objective of mitigating CO_2 emissions. The Paris Agreement proposed limiting global temperature increases to 2°C over pre-industrial times (Wang *et al.*, 2023c). At the 26th United Nations Climate Change Conference (COP 26) in 2021, governments reasserted their carbon reduction expectations, committing to carbon neutrality by mid-century (Chen *et al.*, 2022). It is clear, however, that achieving these goals remains a challenge, and urgent action is required, including greater clean technology innovation. Carbon capture, utilization, and storage (CCUS) is one such crucial clean technology (Wang *et al.*, 2021b) in that, as its name suggests, it captures, transports, and stores CO_2 emissions, allowing the accomplishment of a "carbon budget," ensuring that climate change goals are met (Fan *et al.*, 2021b). CCUS is necessary for reducing emissions from fossil fuel-fired power plants and industrial manufacturing. Thus, it is considered the only technological option for decarbonizing fossil fuel energy generation (Sanchez *et al.*, 2018). Data from the China CCUS Annual Report (2021) reveal that CCUS has the potential to reduce emissions, on the order of 0.6–2.1 $GtCO_2$, helping China meet its carbon neutrality targets.

CCUS has yet to be widely adopted, however, despite its potential to reduce emissions, due to its high cost, which remains a significant obstacle to commercialization. In China, CCUS is still in the R&D stage, and retrofitting CCUS presents significant costs and other risks, particularly in developing countries (Fan *et al.*, 2022). Installing carbon capture devices in fossil fuel power plants can increase costs by 0.26–0.4 yuan/kWh, making it unappealing for companies without government incentives (Yao *et al.*, 2018). Thus, CCUS, despite its emissions reduction potential, lacks commercial application and economic development value, contributing to its ambiguous impact on green total factor productivity (GTFP), as described hereinafter. As mentioned earlier, the high cost of CCUS impedes sustainable development, making it a burden on society. Conversely, innovation is crucial for green development, especially of such end-of-pipe treatment technologies as CCUS, which can effectively promote green development and achieve Chinese carbon neutrality (Li *et al.*, 2022a). The Korean government has accordingly marked CCUS as one of its "27 Green Technologies for Sustainable Economic Growth" (Jung *et al.*, 2022). CCUS was also designated in the US–China Joint Statement Addressing the Climate Crisis, again, signifying its importance in emissions reduction and green development.

Previous studies have focused on investigating the impact of CCUS on carbon emissions (Greig and Uden, 2021; Tapia *et al.*, 2018), cost optimization (Fan *et al.*, 2022), or carbon neutrality (Li *et al.*, 2022b), while the impact of China's CCUS development on GTFP has received little scholarly attention. It is also important to example the current state of CCUS innovation in China, which provides a basis for further exploration of its social and economic effects. Accordingly, in this chapter, we first assess the state of CCUS innovation in China, then its impact on GTFP. We also tease out the moderating role of environmental regulation and the mediating roles of industrial infrastructure and carbon emissions efficiency. Finally, we employ the panel quantile model to show that CCUS innovation and GTFP have a nonlinear relationship.

This chapter contributes to the literature as follows. First, it assesses the impact of CCUS innovation on GTFP, which facilitates the strategic choice of CCUS deployment in China and the government's achievement of sustainable development goals. Second, we explore the intrinsic mechanisms whereby CCUS affects GTFP, which, again, benefits government CCUS initiatives more conducive to faster green development. Third, we examine the moderating role of environmental regulation, finding that it increases the positive impact of CCUS on GTFP. Finally, we explore the asymmetric impact of CCUS innovation on GTFP, finding that it is more prominent in areas with higher levels of GTFP, which is crucial for governments making policies according to local circumstances.

The rest of this chapter is arranged as follows. Section 2 summarizes the relevant literature and cites possible research gaps. Section 3 presents estimated models and data. Section 4 shows results and initial discussion. Section 5 continues the discussion, and Section 6 provides a conclusion.

2. Literature Review

2.1. *The GTFP-technology nexus*

Many scholars have studied GTFP, some of them linking it with innovation and green technology. Tong *et al.* (2022) investigate the role of green technology and green finance in promoting GTFP, showing that the former increases the latter, consistent with Xie *et al.* (2021). Song *et al.* (2022) bring a more narrowed focus to investigating the impact of green technology on GTFP based on enterprise-level data, finding that it also contributes to business GTFP, and that labor productivity enhancement and environmental regulation are the impact mechanisms. Based on a dataset of 28 Chinese provinces during 2011–2021, Jiakui *et al.* (2023) also find that green technology innovation drives GTFP higher. Du and Li (2019) demonstrate the significance of green technology innovation in raising GTFP. Similar conclusions can be found in Chen *et al.* (2023), Cheng *et al.* (2023), Lyu *et al.* (2023), Tian and Pang (2022), and Wang *et al.* (2023b). While Wang *et al.* (2021a) support the above findings, they emphasize that although innovation can stimulate GTFP in Chinese provinces, such innovation has a significantly negative spatial spillover effect on GTFP, indicating that such innovation inhibits GTFP in surrounding provinces. Fang *et al.* (2021) show a negative relationship between technology and GTFP by stating that technology and innovation investment inhibit GTFP, while Wu and Zhang (2020) find that although IT can positively affect GTFP in the short term, its long-term impacts are blunted.

2.2. *Literature gaps*

In recent years, a growing body of scholars has shed light on CCUS, particularly the technical aspects. Lin *et al.* (2022) discuss chemical absorption technologies, while Zhang *et al.* (2013) shed light on the

CCUS roadmap. Lin and Tan (2021) studied CCUS development in terms of enhancing oil recovery (EOR). Li *et al.* (2022b) investigated specific CO_2 emissions reductions that could be expected from CCUS in coal-fired power plants, backed up by Zheng *et al.* (2022). Wei *et al.* (2022) similarly investigated CCUS in the steel industry. Sun *et al.* (2018) documented CCUS CO_2 storage capacities. Yao *et al.* (2018) focused on the business performance of CCUS projects. Li *et al.* (2022c) summarized the demonstration and innovation of CCUS by China Petroleum & Chemical Corporation (Sinopec). Chen and Wu (2022) mentioned CCUS operation and management. Hasan *et al.* (2022) and Jiang and Ashworth (2021) show evaluations CCUS projects and policies. Wei *et al.* (2021) comprehensively analyzed CCUS projects' strengths, weaknesses, opportunities, and threats. Jiang *et al.* (2020) systematically reviewed current CCUS policies, pointed out some of their shortcomings. Jiang *et al.* (2022) focused on how the press and social media discuss the CCUS value chain.

This analysis suggests that there are gaps in the literature. First, there is no consensus on the relationship between GTFP and technology. While some research suggests that innovation can stimulate GTFP, other studies find a negative correlation. While the current value of CCUS innovation in low-carbon technology terms is significant, the aforementioned lack of research on the relationship between CCUS innovation and GTFP remains and is ripe for investigation. Such research on CCUS also fails to measure its socioeconomic value, nor does it explore the aforementioned asymmetric relation between CCUS innovation and GTFP.

3. Methodology and Data

3.1. *Estimation model*

Based on the foregoing, we believe CCUS innovation may have an impact on GTFP, and we have constructed a theoretical model to evaluate this impact within an environmental impact theory context, combined with factors affecting GTFP, including economy, population, trade, and fiscal expenditure.

$$gtfp_{it} = f(ccus_{it}, pgdp_{it}, urb_{it}, fdi_{it}, gov_{it}) \qquad (1)$$

where *i* denotes province, *t* denotes year. *gtfp* is green total factor productivity (GTFP), *ccus* is CCUS innovation index, *urb* is urbanization level, *fdi* is foreign direct investment, and *gov* is general government expenditures.

To remove heteroscedasticity and facilitate estimating the magnitude of the effect, this study takes a logarithmic transformation of Equation (1). It also employs a dynamic estimation model taking into account endogeneity and the time lag effect of environmental variables, yielding Equation (2):

$$lngtfp_{it} = \delta_0 + \delta_1 lngtfp_{i,t-1} + \delta_2 lnccus_{it} + \sum_{k=3}^{6} \delta_k X_{it} + \varepsilon_{it} \tag{2}$$

where δ_0 is a constant, $\delta_1,\ldots, \delta_6$ are estimated coefficients, ε_{it} is a random error term, $lngtfp_{i,t-1}$ is the first-order lag term of $lngtfp_{it}$, and X_{it} is a set of control variables, i.e., $lnpgdp_{it}$, $lnurb_{it}$, $infdi_{it}$, $lngov_{it}$.

We also ran a mediation test to explore mechanisms of the impact of CCUS innovation on GTFP, using selected estimation models given as Equations (3) and (4):

$$lnM_{it} = \eta_0 + \eta_1 lnM_{i,t-1} + \eta_2 lnccus_{it} + \sum_{k=3}^{6} \eta_k X_{it} + \varepsilon_{it} \tag{3}$$

$$lngtfp_{it} = \lambda_0 + \lambda_1 lngtfp_{i,t-1} + \lambda_2 lnccus_{it} + \lambda_3 lnM_{it} + \sum_{k=4}^{7} \lambda_k X_{it} + \varepsilon_{it} \tag{4}$$

where η_0 and λ_0 are constants, $\eta_1,\ldots,\eta_6$ and $\lambda_1,\ldots,\lambda_6$ are estimated coefficients, ε_{it} is the random error term, and M is the mediator, including industrial infrastructure upgrades (*ind*) and carbon emission efficiency (*cee*). We are primarily interested in the coefficient of η_2 in Equation (3), as a significant value here indicates that CCUS innovation can effectively influence the mediating variable. Similarly, a significant coefficient of λ_3 in Equation (4) indicates that the mediating variable can effectively influence GTFP.

We then evaluated the moderating effects influencing the relationship between CCUS innovation and GTFP, with an estimation equation given as Equation (5):

$$lngtfp_{it} = \gamma_0 + \gamma_1 lngtfp_{i,t-1} + \gamma_2 lnccus_{it}$$
$$+ \gamma_3 lnccus_{it} * lner_{it} + \gamma_4 lner_{it} + \sum_{k=5}^{8} \gamma_k X_{it} + \varepsilon_{it} \tag{5}$$

where γ_0 is a constant term, $\gamma_1, \ldots, \gamma_8$ are estimated coefficients, *er* represents the environmental regulation level. We are primarily interested in the coefficient of γ_3, where $\gamma_3 > 0$ indicates that environmental regulation can positively moderate the relationship between CCUS innovation and GTFP and a non-positive value the opposite.

3.2. *Variable measures and data sources*

3.2.1. *Dependent variable*

GTFP is the dependent variable. Data envelopment analysis (DEA) models are widely used due to their utility in evaluating environmental efficiency. We use a non-parametric linear programming method to evaluate the effectiveness of comparable production units, i.e., decision-making units (DMUs), specifically by calculating the ratio of total outputs to total inputs for each DMU. Tone (2001) proposed a method based on slack-based measure (SBM), where the slack variables in the SBM model are directly added to the objective function, and the model is also non-radial, versus traditional DEA models. Tone and Sahoo (2003) considered the importance of adverse outputs such as wastewater and exhaust gas in production in efficiency evaluation systems, establishing another method to measure the efficiency of such adverse outputs based on the SBM model. Most efficiency evaluations show, however, that multiple DMUs have 100% efficiency (Li and Shi, 2014). Accordingly, this study will adopt the super-efficiency SBM-DEA model with adverse outputs, to improve our ability to screen multiple effective DMUs (Lee, 2021; Zhou *et al.*, 2019).

Efficiency changes in the DEA model over time are calculated chiefly using the Malmquist-Luenberger (ML) index, based on the directional distance function proposed by Chung *et al.* (1997), and the Global ML (GML) index measured by Oh (2010). Given that the latter has transitivity and cyclical accumulation that the former lacks (Zhuo *et al.*, 2022), this study uses the GML index to measure GTFP by province. The input variables are labor, investment, and energy, represented by year-end numbers of employees, fixed assets, and total energy consumption, respectively. Fixed assets are calculated using the perpetual inventory method with 2006 the designated base year. The expected output is GDP, which is processed at constant 2006 prices. Non-expected outputs are CO_2 and SO_2 emissions, as well as chemical oxygen demand emissions. Figure 1 shows GTFP changes in 30 provinces.

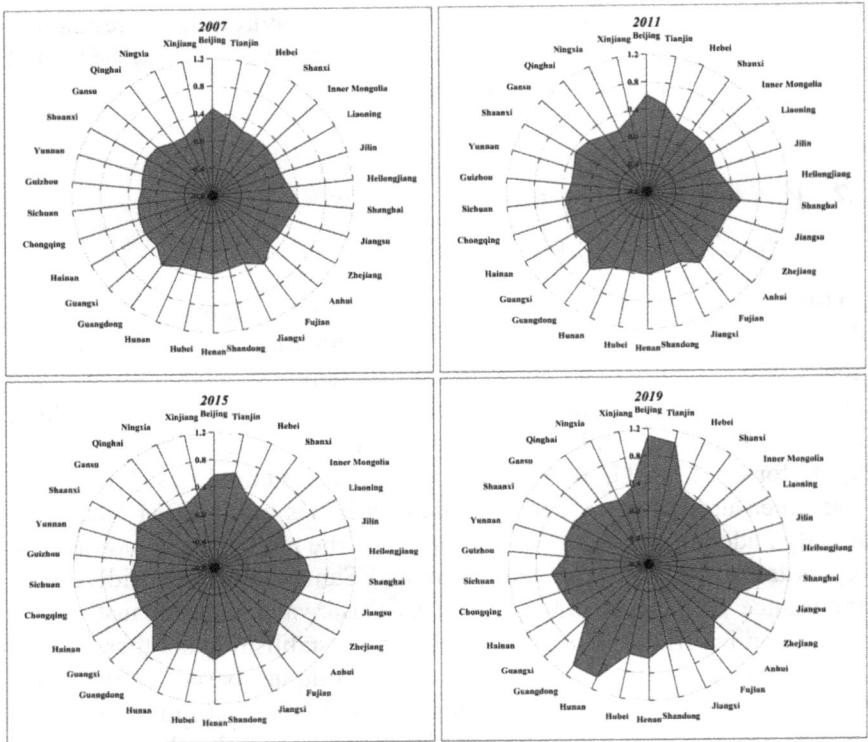

Figure 1.   GTFP across 30 Chinese provinces in 2007, 2011, 2015, and 2019.

3.2.2. *Core independent variable*

The core independent variable in this chapter is the CCUS innovation index. The study retrieved the number of CCUS patents (*ccusc*) from the China National Intellectual Property Administration (CNIPA) database using keyword and patent code information. The CCUS innovation index was then calculated based on the Perpetual Inventory Method (Cheng and Yao, 2021):

$$ccus_{it} = ccusc_{it} + (1 - \rho)ccus_{i,t-1} \tag{6}$$

where ρ denotes depreciation rate, which is set to 10 percent.

3.2.3. *Control variables*

Drawing on prior research, this study selects the following control variables as having potential to affect changes in GTFP:

(1) Economic growth (*pgdp*), which has a dual effect on green efficiency: While it may drive more resource consumption and thus have a negative impact on the environment, it may also bring better governance and technological levels, improving environmental governance (Sun *et al.*, 2022). We use per capita GDP as a proxy.
(2) Urbanization (*urb*): This usually significantly affects the environment and is also characteristic of China's urban–rural divide. This study signifies this with the proportion of urban population to total population (Zhao *et al.*, 2022b).
(3) Foreign direct investment (*fdi*): In addition to being a key driver of China's economic growth, this is also an important determinant of environmental change. While FDI may be in carbon-intensive industries, leading to increased pollution, it may also be in environmentally friendly enterprises, accelerating technological upgrades and thus improving the environment. This study measures FDI with a ratio of total imports by businesses with FDI to GDP.
(4) General budget expenditures of local governments (*gov*): We include this variable because, while government financial support may help business use more funds for green development, it may also drive increased pollution through business pursuit of profit (Ren, 2020).

3.2.4. *Mediating variables*

This study uses the mediating variables of industrial infrastructure upgrades (*lnind*) and carbon emissions efficiency (*lncee*). The former represents changes in the proportions of Chinese industry, thereby objectively reflecting development trends in local sectors. Per Li *et al.* (2019), this research uses the proportion of tertiary industry value added to secondary industry value added to represent industrial infrastructure upgrades. The latter represents synergies between carbon emissions and economic development. This chapter also uses the aforementioned GML index to calculate carbon emissions efficiency by province, with the only non-desired output thereof being CO_2 emissions.

3.2.5. *Moderating variable*

In this chapter, environmental regulation index comprises per capita income, education, population density, and age, per Pargal and Wheeler (1996). It is based on the rationale that affluent, well-educated residents tend to have greater bargaining power in environmental governance negotiations. As population density increases, individuals may be exposed to more pollution, which may raise awareness of environmental concerns. Younger population segments also may be more susceptible to pollution (Zhao *et al.*, 2022a). The proxies are as follows: For per capita income, average urban and rural wages; for education, the proportion of the population with at least a high school diploma; for population density, the ratio of urban population to urban area; and for age, the proportion of the population under 15 years of age.

3.3. *Data sources*

As mentioned above, this chapter evaluates the impact of CCUS innovation on GTFP, based on calculations of CCUS innovation index and GTFP using panel data from 30 Chinese provinces in 2007–2019. CO_2 emissions data used in GTFP calculations are sourced from Carbon Emission Accounts and Datasets (CEADs) (Guan *et al.*, 2021; Shan *et al.*, 2018; Shan *et al.*, 2020; Shan *et al.*, 2016), with other data obtained from the China Statistical Yearbook (NBS, 2022). CCUS patent data are manually compiled by the authors. Table 1 shows descriptive statistics for all variables.

Table 1.   Descriptive statistics of variables (log).

Variable	Mean	Std. Dev.	Min.	Median	Max.
lngtfp	−0.9889	0.3586	−1.9695	−1.0060	0.0879
lnccus	1.6427	0.5824	0.0000	1.6745	3.2097
lnpgdp	10.5395	0.5623	8.9591	10.5473	12.0111
lnurb	−0.6208	0.2337	−1.2641	−0.6239	−0.0603
lnfdi	10.9396	1.4170	7.6885	10.8539	14.4850
lngov	8.0324	0.7523	5.4883	8.1408	9.7583
lnind	0.0942	0.3872	−0.6405	0.0493	1.6552
lneef	0.1735	0.5550	−1.2546	0.2086	1.5719
lner	−1.1069	0.3197	−2.2039	−1.0671	−0.3905

Notes: Mean: average value; Std. Dev.: standard deviation; Min, Median, and Max: minima, median, and maxima, respectively.

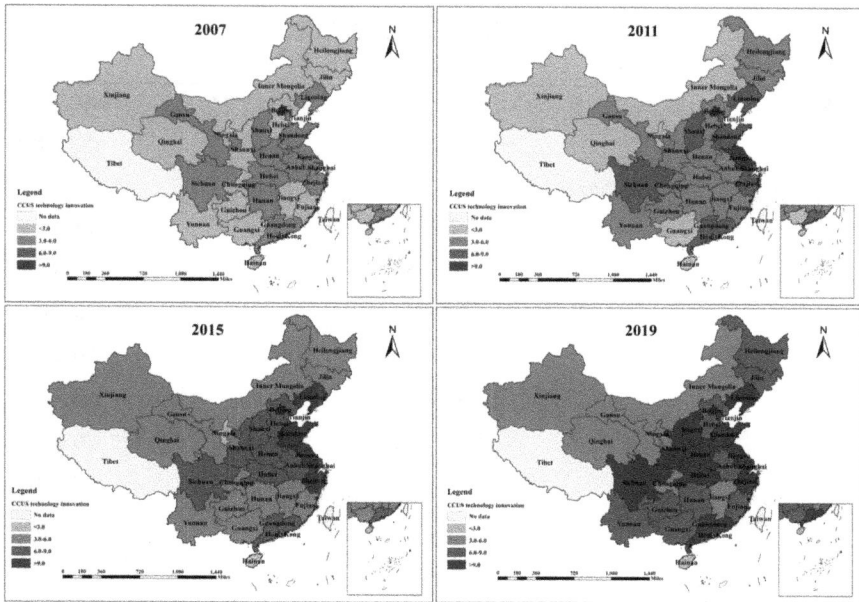

Figure 2. CCUS innovation indices in 30 Chinese provinces for 2007, 2011, 2015, and 2019.

4. Results and Discussion

4.1. *The characteristics of CCUS innovation*

Figure 2 shows CCUS innovation levels in 30 Chinese provinces in 2007, 2011, 2015, and 2019. The darker the color, the higher the level of CCUS innovation. Overall, China's CCUS innovation level has significantly improved, which is attributed to increased government attention to CCUS. On a per province level, we find that provinces with higher CCUS innovation levels are concentrated in coastal areas, including Beijing and Jiangsu, implying their relatively more robust R&D capabilities, which are more conducive to rapid growth in CCUS-related patents.

4.2. *Benchmark regression analysis*

Table 2 shows our benchmark regression analysis of the impact of CCUS innovation on GTFP. Columns 1 and 2 show fixed effects model results, first

Table 2.   Benchmark regression results on GTFP and CCUS innovation.

Variable	(1)	(2)	(3)	(4)
$lngtfp_{i,t-1}$			0.912***	0.928***
			(244.77)	(28.87)
$lnccus$	0.544***	0.153***	0.042***	0.083***
	(31.39)	(2.93)	(13.60)	(3.44)
$lnpgdp$		−0.045		0.018
		(−0.60)		(0.58)
$lnurb$		0.142		−0.052
		(1.146)		(−1.11)
$lnfdi$		0.070***		−0.003
		(3.78)		(−0.19)
$lngov$		0.185***		−0.038*
		(3.92)		(−1.73)
Constant	−1.883***	−2.93***	−0.111***	−0.046
	(−64.97)	(−5.25)	(−13.27)	(−0.16)
AR(1)			0.032	0.028
AR(2)			0.100	0.100
Hansen			0.860	0.367

Notes: ***, **, and * indicate statistical significance at $p = 0.01$, $p = 0.05$, and $p = 0.1$, respectively. Values in parentheses represent t statistics.

without, then with, control variables. Columns 3 and 4 shows system GMM estimation dynamic model results, again first without, then with, control variables. Pursuant to dynamic model prerequisites, AR(1) p-values are all less than 0.1, while AR(2) p-values are all greater than 0.1, indicating first-order, but not second-order, autocorrelation in the GMM model data. The Hansen test gives p-value greater than 0.1, indicating that the instrumental variables selected for the dynamic GMM model are valid. The estimated coefficients of the four key variables in Table 2 are all significantly positive, indicating that CCUS innovation can effectively elevate China's GTFP level, as supported by Li *et al.* (2022a). The explanation is as follows. CCUS can capture, store, or reuse CO_2 generated in industry and energy production (Li *et al.*, 2016; Yang *et al.*, 2022), thereby effectively reducing greenhouse gas emissions (GHGs). It can also reduce reliance on imported oil and gas, lower energy costs, increase employment opportunities, and promote clean industry development, thereby improving China's GTFP (Mennicken *et al.*, 2016). High costs are the main barrier to CCUS deployment. Combined with climate change and carbon neutrality goals, the expense drives companies to rapidly adopt renewable energy and other clean energy sources to achieve green economic transformation (Fan *et al.*, 2021a).

4.3. *Robustness checks*

This chapter employs the following robustness tests. First, we re-estimate Equation (2) using Difference GMM, which performs better than System GMM in the presence of heteroscedasticity and autocorrelation, providing relatively more robust estimates. Table 3, Column (1) shows the results.

Table 3.   Robustness test results.

Variable	(1)	(2)
lngtfp$_{i,t-1}$	1.184***	0.103***
	(23.51)	(3.08)
lnccus	0.055**	0.124***
	(2.36)	(5.11)
lnpgdp	0.040	0.000
	(0.83)	(0.07)
lnurb	0.008	−0.004
	(0.14)	(−0.28)
lnfdi	−0.046***	−0.001
	(−3.53)	(−0.37)
lngov	−0.100***	−0.005*
	(−3.53)	(−1.94)
Constant	1.051***	0.065
	(3.16)	(1.22)
AR(1)	0.039	0.044
AR(2)	0.104	0.105
Hansen	0.792	0.502

Notes: *** and * indicate statistical significance at $p = 0.01$ and $p = 0.1$, respectively. Values in parentheses represent *t* statistics.

Second, pursuant to Wang *et al.* (2023a), we recalculate the CCUS innovation index using Equation (7):

$$ccus_{it} = \sum_{j=0}^{t} ccusc_{ij} exp[-\omega_1(t-j)]\{1 - exp[-\omega_2(t-j)]\} \tag{7}$$

where ω_1 represents the depreciation rate and ω_2 the diffusion rate. Then we re-estimate Equation (2) with the new CCUS innovation index, with results as per Table 3, Column (2). The coefficients of *lnccus* are significantly positive in both columns, indicating that the benchmark regression results are robust.

5. Further Discussion

The following discussion focus on a more in-depth exploration of the relationship between CCUS innovation and GTFP. Section 5.1 investigates transmission pathways between these. Section 5.2 examines the moderating effect of environmental regulations. Section 5.3 studies their asymmetric relationship.

5.1. *Mediating effects between CCUS innovation and GTFP*

Table 4 shows estimates of mediating effects. Columns (1) and (2) respectively show estimates of industrial infrastructure upgrade mediating effects in Model (3) and Model (4). In Column (1), the coefficient of *lnccus* is significantly positive, indicating that CCUS innovation can effectively promote such upgrades. In Column (2), the coefficient of *lnind* is also significantly positive, indicating that these can effectively increase GTFP. These confirm that industrial infrastructure upgrades significantly mediate the impact of CCUS innovation on GTFP. Chinese companies have gradually realized the importance of CCUS and are using it to reduce carbon emissions, improving environmental quality while reducing energy costs and enhancing competitiveness. This effectively promotes gradually upgrading China's industrial infrastructure, transforming to low-carbon, environmentally friendly manufacturing methods. Conversely, CCUS also promotes China's transformation to cleaner industry, by driving clean energy development and assisting with solar and wind energy storage, thereby gradually reducing fossil

Table 4. Mediating role of industrial infrastructure upgrades and carbon emissions efficiency.

Variable	*lnind* (1)	*lngtfp* (2)	*lncee* (3)	*lngtfp* (4)
$lngtfp_{i,t-1}$		0.928***		0.657***
		(40.07)		(9.76)
$lnind_{i,t-1}$	0.901***			
	(20.89)			
$lncee_{i,t-1}$			0.962***	
			(39.55)	
lnccus	0.098**	0.047*	0.041**	0.052***
	(2.65)	(1.89)	(2.06)	(5.22)
lnind		0.074***		
		(4.39)		
lncee				0.394***
				(4.78)
lnpgdp	−0.284***	0.065**	0.057**	−0.011
	(−11.64)	(1.89)	(2.16)	(−0.43)
lnurb	0.468***	−0.141***	−0.074**	−0.076
	(6.77)	(2.30)	(−2.37)	(−1.61)
lnfdi	0.005	−0.009	−0.017	0.034***
	(0.32)	(−1.00)	(−1.58)	(3.95)
lngov	0.118***	−0.056***	−0.052***	−0.056***
	(5.56)	(−2.86)	(−3.00)	(−3.89)
Constant	2.163***	−0.341	−0.055	−0.380*
	(9.56)	(−1.56)	(−0.32)	(−1.96)
AR(1)	0.006	0.035	0.000	0.077
AR(2)	0.138	0.100	0.316	0.119
Hansen	0.207	0.350	0.633	0.854

Notes: ***, **, and * indicate statistical significance at the 1%, 5%, and 10% levels, respectively; the values in parentheses represent *t* statistics.

fuel dependency. Industrial infrastructure contributes more than 70 percent to energy conservation due to its link between manufacturing factors and environmental impact (Zhu *et al.*, 2022). Businesses can improve production tools and increase productivity through industrial infrastructure upgrades, thereby promoting sustained economic growth (Dong *et al.*, 2022). Prior studies also recognize its significance to GTFP (Feng *et al.*, 2019; Liu *et al.*, 2022).

Columns (3) and (4) show the estimates of the carbon emissions efficiency mediation effect in models (3) and (4), respectively. The positive and significant coefficient of *lnccus* in Column (3) indicates that CCUS innovation can effectively improve carbon emissions efficiency. The positive and significant coefficient of *lncee* in Column (4) also suggests that carbon emissions efficiency can increase GTFP. These results confirm the positive transmission effect of carbon emissions efficiency between CCUS innovation and GTFP. CCUS can convert captured CO_2 into other substances, such as synthetic fuels, chemicals, or concrete, which can replace traditional fossil fuels and chemicals, thereby improving the utilization efficiency of carbon emissions (Cormos *et al.*, 2018; Farfan *et al.*, 2019). Conversely reduces CO_2 emissions while promoting economic growth through traditional energy, promoting sustainable industrial development, effectively reducing said utilization efficiency. For example, applying CCUS to coal-fired power plants can increase energy conversion efficiency, reducing fuel consumption and CO_2 emissions (Fan *et al.*, 2020; Fan *et al.*, 2018). Improvement of carbon emissions efficiency promotes rapid GTFP growth through increased economic output per unit energy consumption with reduced carbon emissions and other adverse economic outputs (Li and Liao, 2022).

5.2. *Moderating effect between CCUS innovation and GTFP*

Table 5 illustrates the potential moderating effect of environmental regulations on the relationship between CCUS innovation and GTFP. Columns (1) and (2) include the moderating terms of CCUS and environmental regulations (*lnccus * lner*) and environmental regulations (*lner*), respectively, indicating that the coefficients of the moderating term and the key independent variable *lnccus* are both significantly positive, suggesting that environmental regulations can indeed increase the impact of CCUS innovation on GTFP, with potential reasons as follows. First, environmental regulations require CCUS companies to adopt more environmentally

Table 5. Moderating role of environmental regulation in GTFP-CCUS innovation nexus.

Variable	(1)	(2)
lngtfp$_{i,t-1}$	0.929***	0.921***
	(34.31)	(28.79)
lnccus	0.085***	0.135***
	(3.74)	(2.95)
lner		−0.055
		(−1.36)
*lnccus * lner*	0.052***	0.077***
	(5.33)	(3.87)
lnpgdp	−0.020	−0.013
	(−0.713)	(−0.44)
lnurb	−0.095**	−0.108**
	(−2.288)	(−2.32)
lnfdi	0.024**	0.020**
	(2.55)	(2.05)
lngov	0.003	−0.003
	(0.174)	(−0.159)
Constant	−0.208	−0.309
	(−0.821)	(−1.05)
AR(1)	0.031	0.032
AR(2)	0.128	0.125
Hansen	0.499	0.416

Notes: *** and ** indicate statistical significance at $p = 0.01$ and $p = 0.05$, respectively. Values in parentheses represent t statistics.

friendly technologies and equipment that meet specified standards, thereby improving GTFP. Second, these regulations also compel companies to improve GTFP by utilizing resources more efficiently, increasing CCUS investment, and reducing costs. Finally, by the foregoing reasons, these regulations promote rapid CCUS industry growth, providing more market opportunities and development space, thus making green manufacturing more efficient (Jiang *et al.*, 2020). Figure 3 vividly illustrates this intrinsic connection between CCUS innovation and GTFP.

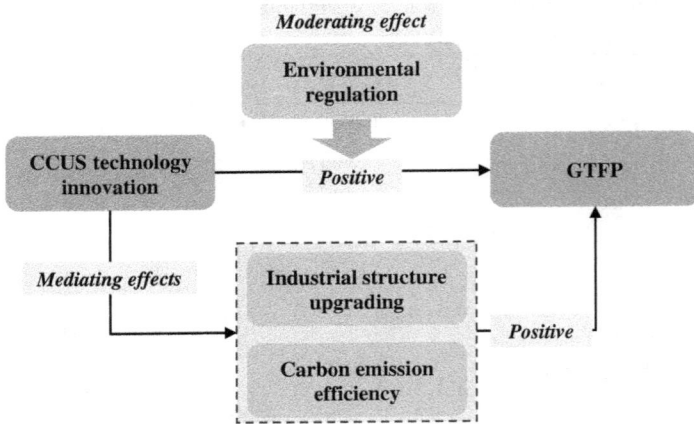

Figure 3.   How CCUS innovation relates to GTFP.

5.3. *Asymmetry check*

Finally, we examine the asymmetric impact of CCUS innovation on GTFP. Table 6 shows the estimates, indicating that while CCUS innovation can effectively promote GTFP improvement in all quantiles, only the coefficient of *lnccus* is significant at the 75th and 90th percentiles. While this indicates that the findings of the impact of CCUS innovation on GTFP are robust, it also suggests that the green growth effect generated by CCUS innovation is more pronounced in regions with higher GTFP. We believe that this is for the following reasons. First, as mentioned above, regions with higher GTFP often have better R&D capabilities, which can better drive green growth by encouraging CCUS research and applications, making CCUS more efficient. Second, such regions typically also have more funding and opportunity, as well as potential for sustainable development, which enhances green development by facilitating more CCUS installations as part of investing in low-carbon and otherwise environmentally friendly industries. Finally, these regions also typically have more rigorous environmental protection policies and regulations, which are devoted to developing environmental protection businesses. Local government will accordingly further promote green growth with policies encouraging CCUS development and commercialization. Figure 4 also shows asymmetry test results for all independent variables, confirming these hypotheses.

Table 6. Asymmetric results of the impact of CCUS technology innovation on GTFP.

Variable	Quantiles				
	10th	25th	50th	75th	90th
lnccus	0.040	0.022	0.018	0.082*	0.147**
	(0.70)	(0.45)	(0.23)	(1.78)	(2.32)
lnpgdp	0.279***	0.203***	0.127	0.247***	0.121*
	(5.21)	(3.96)	(0.90)	(4.32)	(1.93)
lnurb	−0.664***	−0.367***	0.054	−0.189	0.319**
	(−6.70)	(−3.33)	(0.17)	(−1.36)	(2.04)
lnfdi	0.050***	0.049**	0.094***	0.123***	0.100***
	(3.23)	(2.56)	(3.03)	(7.84)	(8.92)
lngov	0.177***	0.174***	0.065	−0.006	0.007
	(4.40)	(5.62)	(0.80)	(−0.16)	(0.15)
Constant	−6.644***	−5.512***	−3.85***	−4.98***	−3.18***
	(−13.13)	(−12.95)	(−2.71)	(−9.20)	(−4.62)

Notes: ***, **, and * indicate statistical significance at the 1%, 5%, and 10% levels, respectively; the values in parentheses represent t statistics.

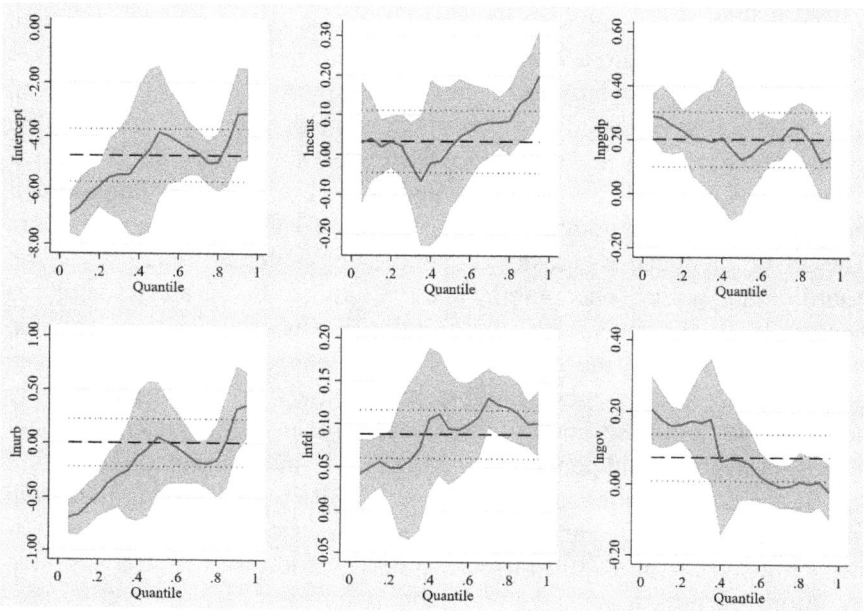

Figure 4. Results of quantile regression on independent variables.

6. Conclusions and Policy Implications

This chapter utilizes the aforementioned panel data from 30 Chinese provinces spanning 2007–2019. The CCUS innovation indices for these provinces are measured by patent searches and provincial GTFP by SBM-GML. We use these results to evaluate the impact of Chinese CCUS innovation on GTFP at the provincial level, and the transmission mechanism, regulatory effects, and asymmetry between them. Our main conclusions are as follows.

First, China has significant CCUS development potential, with innovation indices showing a clear upward trend during the period in question, and developed coastal provinces highest of all. Second, CCUS innovation can effectively drive GTFP higher as well, a finding that robustness tests reinforce. CCUS can also indirectly increase GTFP by promoting industrial infrastructure upgrading and improving carbon emissions efficiency. Environmental regulation also significantly increases the impact of CCUS innovation on GTFP, and CCUS innovation increases GTFP more significantly in regions with more green growth.

The following policy implications are accordingly based on the foregoing. First, CCUS innovation can effectively increase China's GTFP, which in turn encourages coordinated Chinese economic development and environmental protection. We recommend that the government promote CCUS innovation by increasing funding for CCUS R&D, including more grants to business. The government must also similarly improve the environmental policy and regulatory framework to support CCUS innovation and commercialization and strengthen supervision of polluting businesses, thereby encouraging better private-sector CCUS implementation. The government can also foster cooperation by better communicating and coordinating policy, encouraging CCUS application in such sectors as energy, steel, and chemicals, and building intensive large-scale application models. Finally, the government should encourage CCUS for green growth by providing more flexible preferential tax policies and other subsidies for Chinese energy firms, encouraging independent private-sector innovation, and generally making CCUS more competitive.

Second, CCUS innovation increases GTFP by encouraging industrial infrastructure upgrades and greater carbon emissions efficiency, suggesting the government should further encourage enterprise CCUS adoption by improved policy guidance and support for high carbon-emitting sectors. To this end, government can also improve formulating and supervising carbon emissions indicators, thus guiding business to improve carbon emissions efficiency by adopting more environmentally friendly and energy-saving methods. The government can also encourage green development by establishing CCUS innovation platforms, encouraging faster CCUS rollouts, more rapid industrial transformations, and reduced carbon emissions.

Third, environmental regulation can effectively enhance the green effects of CCUS innovation. The government can accordingly ensure green CCUS development by enacting relevant laws and regulations that clarify CCUS scope and improve its management. The government can also guarantee greater CCUS green growth by establishing an application evaluation mechanism to identify and evaluate environmental and economic benefits from CCUS and implementing public–private supervision.

Finally, CCUS has more significant green economic benefits at higher percentiles, guiding the public and private sectors alike to strengthen CCUS deployment in regions with advanced green economic development, for example, by strengthening CCUS-related industry planning and subsidies to maximize said benefits. Conversely, the government must adopt less expensive paths to green development for regions trailing in this regard. Measures might include accelerating renewable energy deployment, improving industry operations, and enhancing environmental protection technologies.

Acknowledgments

This chapter was sponsored by the National Social Science Foundation of China (Grant No. 20VGQ003). All errors remain the authors' sole responsibility. This chapter is a modified version of a forthcoming paper in Asian Economic Papers.

Data Availability

Data in support of our findings are available from the corresponding author upon reasonable request.

Disclosure

No potential conflict of interest was reported by the authors.

References

Chen, C., Ye, F., Xiao, H., Xie, W., Liu, B., and Wang, L. (2023). The digital economy, spatial spillovers and forestry green total factor productivity. *Journal of Cleaner Production, 405*, 136890.

Chen, L., Msigwa, G., Yang, M., Osman, A.I., Fawzy, S., Rooney, D.W., and Yap, P.-S. (2022). Strategies to achieve a carbon neutral society: A review. *Environmental Chemistry Letters, 20*, 2277–2310.

Chen, X. and Wu, X. (2022). The roles of carbon capture, utilization and storage in the transition to a low-carbon energy system using a stochastic optimal scheduling approach. *Journal of Cleaner Production, 366*, 132860.

Cheng, Y., Lv, K., and Zhu, S. (2023). How does digital financial inclusion promote green total factor productivity in China? An empirical analysis from the perspectives of innovation and entrepreneurship. *Process Safety and Environmental Protection, 174*, 403–413.

Cheng, Y. and Yao, X. (2021). Carbon intensity reduction assessment of renewable energy technology innovation in China: A panel data model with cross-section dependence and slope heterogeneity. *Renewable & Sustainable Energy Reviews, 135*, 110157.

Chung, Y. H., Färe, R., and Grosskopf, S. (1997). Productivity and undesirable outputs: A directional distance function approach. *Journal of Environmental Management, 51*, 229–240.

Cormos, A.-M., Dinca, C., Petrescu, L., Andreea Chisalita, D., Szima, S., and Cormos, C.-C. (2018). Carbon capture and utilisation technologies applied to energy conversion systems and other energy-intensive industrial applications. *Fuel, 211*, 883–890.

Dong, X., Chen, Y., Zhuang, Q., Yang, Y., and Zhao, X. (2022). Agglomeration of productive services, industrial structure upgrading and green total factor productivity: An empirical analysis based on 68 prefectural-level-and-above cities in the Yellow River Basin of China. *International Journal of Environmental Research and Public Health, 19*(18), 11643.

Du, K. and Li, J. (2019). Towards a green world: How do green technology innovations affect total-factor carbon productivity. *Energy Policy, 131*, 240–250.

Fan, J.-L., Shen, S., Wei, S.-J., Xu, M., and Zhang, X. (2020). Near-term CO_2 storage potential for coal-fired power plants in China: A county-level source-sink matching assessment. *Applied Energy, 279*, 115878.

Fan, J.-L., Xu, M., Li, F., Yang, L., and Zhang, X. (2018). Carbon capture and storage (CCS) retrofit potential of coal-fired power plants in China: The technology lock-in and cost optimization perspective. *Applied Energy, 229*, 326–334.

Fan, J.-L., Xu, M., Wei, S., Shen, S., Diao, Y., and Zhang, X. (2021a). Carbon reduction potential of China's coal-fired power plants based on a CCUS source-sink matching model. *Resources, Conservation and Recycling, 168,* 105320.

Fan, J., Li, Z., Li, K., and Zhang, X. (2022). Modelling plant-level abatement costs and effects of incentive policies for coal-fired power generation retrofitted with CCUS. *Energy Policy, 165,* 112959.

Fan, J., Xu, M., Wei, S., Shen, S., Diao, Y., and Zhang, X. (2021b). Carbon reduction potential of China's coal-fired power plants based on a CCUS source-sink matching model. *Resources, Conservation and Recycling, 168,* 105320.

Fang, C., Cheng, J., Zhu, Y., Chen, J., and Peng, X. (2021). Green total factor productivity of extractive industries in China: An explanation from technology heterogeneity. *Resources Policy, 70,* 101933.

Farfan, J., Fasihi, M., and Breyer, C. (2019). Trends in the global cement industry and opportunities for long-term sustainable CCU potential for Power-to-X. *Journal of Cleaner Production, 217,* 821–835.

Feng, Y., Zhong, S., Li, Q., Zhao, X., and Dong, X. (2019). Ecological well-being performance growth in China (1994–2014): From perspectives of industrial structure green adjustment and green total factor productivity. *Journal of Cleaner Production, 236,* 117556.

Greig, C. and Uden, S. (2021). The value of CCUS in transitions to net-zero emissions. *The Electricity Journal, 34,* 107004.

Guan, Y., Shan, Y., Huang, Q., Chen, H., Wang, D., and Hubacek, K. (2021). Assessment to China's recent emission pattern shifts. *Earth's Future, 9,* e2021EF002241.

Hasan, M. M. F., Zantye, M. S., and Kazi, M.-K. (2022). Challenges and opportunities in carbon capture, utilization and storage: A process systems engineering perspective. *Computers & Chemical Engineering, 166,* 107925.

Jiakui, C., Abbas, J., Najam, H., Liu, J., and Abbas, J. (2023). Green technological innovation, green finance, and financial development and their role in green total factor productivity: Empirical insights from China. *Journal of Cleaner Production, 382,* 135131.

Jiang, K. and Ashworth, P. (2021). The development of carbon capture utilization and storage (CCUS) research in China: A bibliometric perspective. *Renewable and Sustainable Energy Reviews, 138,* 110521.

Jiang, K., Ashworth, P., Zhang, S., and Hu, G. (2022). Print media representations of carbon capture utilization and storage (CCUS) technology in China. *Renewable and Sustainable Energy Reviews, 155,* 111938.

Jiang, K., Ashworth, P., Zhang, S., Liang, X., Sun, Y., and Angus, D. (2020). China's carbon capture, utilization and storage (CCUS) policy: A critical review. *Renewable and Sustainable Energy Reviews, 119,* 109601.

Jung, S.-h., Kim, H., Kang, Y., and Jeong, E. (2022). Analysis of Korea's green technology policy and investment trends for the realization of carbon neutrality: Focusing on CCUS technology. *Processes, 10,* 501.

Lee, H.-S. (2021). An integrated model for SBM and Super-SBM DEA models. *Journal of the Operational Research Society, 72,* 1174–1182.

Li, G. and Liao, F. (2022). Input digitalization and green total factor productivity under the constraint of carbon emissions. *Journal of Cleaner Production, 377,* 134403.

Li, H. and Shi, J. (2014). Energy efficiency analysis on Chinese industrial sectors: An improved Super-SBM model with undesirable outputs. *Journal of Cleaner Production, 65,* 97–107.

Li, J., Dong, K., and Dong, X. (2022a). Green energy as a new determinant of green growth in China: The role of green technological innovation. *Energy Economics, 114,* 106260.

Li, K., Shen, S., Fan, J., Xu, M., and Zhang, X. (2022b). The role of carbon capture, utilization and storage in realizing China's carbon neutrality: A source-sink matching analysis for existing coal-fired power plants. *Resources, Conservation and Recycling, 178,* 106070.

Li, Q., Chen, Z. A., Zhang, J. T., Liu, L. C., Li, X. C., and Jia, L. (2016). Positioning and revision of CCUS technology development in China. *International Journal of Greenhouse Gas Control, 46,* 282–293.

Li, T., Li, Y., An, D., Han, Y., Xu, S., Lu, Z., and Crittenden, J. (2019). Mining of the association rules between industrialization level and air quality to inform high-quality development in China. *Journal of Environmental Management, 246,* 564–574.

Li, Y., Rui, W., Qingmin, Z., and Zhaojie, X. (2022c). Technological advancement and industrialization path of Sinopec in carbon capture, utilization and storage, China. *Energy Geoscience,* 100107.

Lin, B. and Tan, Z. (2021). How much impact will low oil price and carbon trading mechanism have on the value of carbon capture utilization and storage (CCUS) project? Analysis based on real option method. *Journal of Cleaner Production, 298,* 126768.

Lin, Q., Zhang, X., Wang, T., Zheng, C., and Gao, X. (2022). Technical perspective of carbon capture, utilization, and storage. *Engineering, 14,* 27–32.

Liu, Y., Yang, Y., Li, H., and Zhong, K. (2022). Digital economy development, industrial structure upgrading and green total factor productivity: Empirical evidence from China's cities. *International Journal of Environmental Research and Public Health, 19*(4), 2414.

Lyu, Y., Wang, W., Wu, Y., and Zhang, J. (2023). How does digital economy affect green total factor productivity? Evidence from China. *Science of The Total Environment, 857,* 159428.

Mennicken, L., Janz, A., and Roth, S. (2016). The German R&D program for CO_2 utilization — Innovations for a green economy. *Environmental Science and Pollution Research, 23,* 11386–11392.

NBS (2022). *China Statistical Yearbook.* Beijing: China Statistics Press.

Oh, D.-h. (2010). A global Malmquist-Luenberger productivity index. *Journal of Productivity Analysis, 34*, 183–197.

Pargal, S. and Wheeler, D. (1996). Informal regulation of industrial pollution in developing countries: Evidence from Indonesia. *Journal of Political Economy, 104*, 1314–1327.

Ren, Y. (2020). Research on the green total factor productivity and its influencing factors based on system GMM model. *Journal of Ambient Intelligence and Humanized Computing, 11*, 3497–3508.

Sanchez, D. L., Johnson, N., McCoy, S. T., Turner, P. A., Mach, K. J. (2018). Near-term deployment of carbon capture and sequestration from biorefineries in the United States. *Proceedings of the National Academy of Sciences, 115*, 4875–4880.

Shan, Y., Guan, D., Zheng, H., Ou, J., Li, Y., Meng, J., Mi, Z., Liu, Z., and Zhang, Q. (2018). China CO_2 emission accounts 1997–2015. *Scientific Data, 5*, 170201.

Shan, Y., Huang, Q., Guan, D., and Hubacek, K. (2020). China CO_2 emission accounts 2016–2017. *Scientific Data, 7*, 54.

Shan, Y., Liu, J., Liu, Z., Xu, X., Shao, S., Wang, P., and Guan, D. (2016). New provincial CO_2 emission inventories in China based on apparent energy consumption data and updated emission factors. *Applied Energy, 184*, 742–750.

Song, M., Peng, L., Shang, Y., and Zhao, X. (2022). Green technology progress and total factor productivity of resource-based enterprises: A perspective of technical compensation of environmental regulation. *Technological Forecasting and Social Change, 174*, 121276.

Sun, L., Dou, H., Li, Z., Hu, Y., and Hao, X. (2018). Assessment of CO_2 storage potential and carbon capture, utilization and storage prospect in China. *Journal of the Energy Institute, 91*, 970–977.

Sun, Y., Tang, Y., and Li, G. (2022). Economic growth targets and green total factor productivity: Evidence from China. *Journal of Environmental Planning and Management*, 1–17.

Tapia, J. F. D., Lee, J., Ooi, R. E. H., Foo, D. C. Y., and Tan, R. R. (2018). A review of optimization and decision-making models for the planning of CO_2 capture, utilization and storage (CCUS) systems. *Sustainable Production and Consumption, 13*, 1–15.

Tian, Y. and Pang, J. (2022). The role of internet development on green total-factor productivity — An empirical analysis based on 109 cities in Yangtze River economic belt. *Journal of Cleaner Production, 378*, 134415.

Tone, K. (2001). A slacks-based measure of efficiency in data envelopment analysis. *European Journal of Operational Research, 130*, 498–509.

Tone, K. and Sahoo, B. K. (2003). Scale, indivisibilities and production function in data envelopment analysis. *International Journal of Production Economics, 84*, 165–192.

Tong, L., Chiappetta Jabbour, C. J., Belgacem, S. b., Najam, H., Abbas, J. (2022). Role of environmental regulations, green finance, and investment in green technologies in green total factor productivity: Empirical evidence from Asian region. *Journal of Cleaner Production, 380*, 134930.

Wang, H., Cui, H., and Zhao, Q. (2021a). Effect of green technology innovation on green total factor productivity in China: Evidence from spatial Durbin model analysis. *Journal of Cleaner Production, 288*, 125624.

Wang, J., Dong, X., and Dong, K. (2023a). Does renewable energy technological innovation matter for green total factor productivity? Empirical evidence from Chinese provinces. *Sustainable Energy Technologies and Assessments, 55*, 102966.

Wang, J., Liu, Y., Wang, W., and Wu, H. (2023b). How does digital transformation drive green total factor productivity? Evidence from Chinese listed enterprises. *Journal of Cleaner Production, 406*, 136954.

Wang, J., Zhao, J., Dong, K., and Dong, X. (2023c). How does the internet economy affect CO_2 emissions? *Evidence from China. Applied Economics, 55*, 447–466.

Wang, N., Akimoto, K., and Nemet, G. F. (2021b). What went wrong? Learning from three decades of carbon capture, utilization and sequestration (CCUS) pilot and demonstration projects. *Energy Policy, 158*, 112546.

Wei, N., Li, X., Liu, S., Lu, S., and Jiao, Z. (2021). A strategic framework for commercialization of carbon capture, geological utilization, and storage technology in China. *International Journal of Greenhouse Gas Control, 110*, 103420.

Wei, N., Liu, S., Jiao, Z., and Li, X. (2022). A possible contribution of carbon capture, geological utilization, and storage in the Chinese crude steel industry for carbon neutrality. *Journal of Cleaner Production, 374*, 133793.

Wu, L. and Zhang, Z. (2020). Impact and threshold effect of Internet technology upgrade on forestry green total factor productivity: Evidence from China. *Journal of Cleaner Production, 271*, 122657.

Xie, F., Zhang, B., and Wang, N. (2021). Non-linear relationship between energy consumption transition and green total factor productivity: A perspective on different technology paths. *Sustainable Production and Consumption, 28*, 91–104.

Yang, Y., Xu, W., Wang, Y., Shen, J., Wang, Y., Geng, Z., Wang, Q., and Zhu, T. (2022). Progress of CCUS technology in the iron and steel industry and the suggestion of the integrated application schemes for China. *Chemical Engineering Journal, 450*, 138438.

Yao, X., Zhong, P., Zhang, X., and Zhu, L. (2018). Business model design for the carbon capture utilization and storage (CCUS) project in China. *Energy Policy, 121*, 519–533.

Zhang, X., Fan, J.-L., and Wei, Y.-M. (2013). Technology roadmap study on carbon capture, utilization and storage in China. *Energy Policy*, *59*, 536–550.

Zhao, C., Dong, K., Wang, K., and Dong, X. (2022a). How does energy trilemma eradication reduce carbon emissions? The role of dual environmental regulation for China. *Energy Economics*, *116*, 106418.

Zhao, X., Nakonieczny, J., Jabeen, F., Shahzad, U., and Jia, W. (2022b). Does green innovation induce green total factor productivity? Novel findings from Chinese city level data. *Technological Forecasting and Social Change*, *185*, 122021.

Zheng, J., Duan, H., and Yuan, Y. (2022). Perspective for China's carbon capture and storage under the Paris agreement climate pledges. *International Journal of Greenhouse Gas Control*, *119*, 103738.

Zhou, Y., Liu, W., Lv, X., Chen, X., and Shen, M. (2019). Investigating interior driving factors and cross-industrial linkages of carbon emission efficiency in China's construction industry: Based on Super-SBM DEA and GVAR model. *Journal of Cleaner Production*, *241*, 118322.

Zhu, X., Zhang, B., and Yuan, H. (2022). Digital economy, industrial structure upgrading and green total factor productivity — Evidence in textile and apparel industry from China. *PLOS ONE*, *17*, e0277259.

Zhuo, C., Xie, Y., Mao, Y., Chen, P., and Li, Y. (2022). Can cross-regional environmental protection promote urban green development: Zero-sum game or win–win choice? *Energy Economics*, *106*, 105803.

Chapter 10

Unlocking Economic Viability: Policies for Encouraging Carbon Capture, Utilization, and Storage

Farhad Taghizadeh-Hesary[*,§], **Han Phoumin**[†,¶],
and Lilu Vandercamme[‡]

*School of Global Studies and Tokai Research
Institute for Environment and Sustainability (TRIES),
Tokai University, Tokyo, Japan*

†*Economic Research for ASEAN and East ASIA,
Jakarta, Indonesia*

‡*School of Economics, Keio University, Japan*

§*farhad@tsc.u-tokai.ac.jp*

¶*han.phoumin@eria.org*

‡*liluvandercamme@keio.jp*

Abstract

Carbon capture, utilization, and storage (CCUS) has the potential to play a crucial role in mitigating climate change by capturing CO_2 emissions from industry and power generation, thereby fostering green development. Utilizing CCUS will impose higher costs on many industries, however, making the private sector less eager to invest in it. Following a review of the literature and policies that some businesses have implemented, this chapter provides practical recommendations for making CCUS economically feasible, including technological advancements, policy support, and market mechanisms.

Keywords: Carbon capture, utilization, and storage (CCUS), green development, carbon-free industry, economic viability.

1. Introduction

Sustainability has become a paramount concern worldwide due to the ecological challenges stemming from uncontrolled economic growth (Ofori *et al.*, 2023). Countries strive to balance economic development and environmental action (Alola *et al.*, 2022). Governments worldwide are implementing diverse environmental safeguards and making efforts to curb harmful emissions. Nevertheless, adherence to environmental regulations is slowing industrial development, particularly in emerging and developing economies (Douglas and Bye, 1925). However, such targeted measures as carbon capture, utilization, and storage (CCUS) incentive policies may encourage industry to transition to methods that generate fewer pollutants and capture and utilize their carbon emissions. These policies provide the foundation for this chapter, which seeks to assess the dynamic impact of state-level environmental regulations on corporate trade credit operations.

External costs arise when economic actors excessively exploit ecosystems for personal gain, potentially leading to market inefficiencies. Consequently, governments employ specialized policy instruments to ensure that environmental concerns do not disrupt market economies, preventing industry collapse (Bithas and Latinopoulos, 2021). Ecological restoration is viewed as a crucial objective that governments achieve by imposing social costs, such as carbon pricing and environmental taxes, on manufacturing, and through innovation, pollution mitigation, and economic oversight (Emblemsvåg, 2022). These strategic tools' effectiveness directly influences manufacturing outcomes, which contribute significantly to CO_2 emissions. Shahbaz *et al.* (2018) support this argument that stringent environmental regulations impede productivity.

Delacámara (2008) hypothesized the adverse effects of such environmental pressures on business. The ecological consequences of commerce are intricately linked with various firm-level decisions (Alola, 2019), among which trade is solidly connected to environmental issues. Forson *et al.* (2017) observed the impact of air pollution on firms' credit strategies, positing that suppliers adjust their credit terms in response to severe air pollution, aligning with stress theory. Such adjustments are attributed to financial constraints, heightened risk to clients, and operational challenges in polluted areas. Mitigating such environmental risks is thus crucial to industry expansion.

These risks contribute to broader social challenges, including limitations on public activities in polluted regions and increased strain on such essential services as healthcare. As mentioned earlier, government intervention strives to prevent market failure and establish specific policy boundaries to ensure environmental stability. Such legislative constraints may, again, cause additional economic challenges due to inadequate industrial development and limited employment opportunities, making it imperative to craft regulations that safeguard the environment while preserving productivity.

Innovative carbon reduction solutions have gained currency, particularly in carbon-intensive industries such as the paper sector, where CCUS holds promise (Dziejarski *et al.*, 2023). CCUS stands out among environmental conservation tools for its potential for sustainable innovation and ecological equilibrium. Notably, CCUS has established itself as a forerunner in the global goal of a carbon-neutral future. The paper industry, notorious for its aforementioned carbon footprint, is ready to incorporate CCUS in a transformative way (Onarheim *et al.*, 2017). While such integration is a step toward sustainability by reducing carbon emissions, fulfilling this promise requires a detailed understanding of the delicate balance between CCUS solutions and developing technology in paper mills (Svensson *et al.*, 2021). This study accordingly examines such CCUS deployment in the paper industry, delving into its feasibility, viability, and complexities.

Our primary goal is consistent with the abstract's range of policy proposals. That is, our research aims to provide concrete policy suggestions for making CCUS economically feasible. We traverse the dynamics among technical improvements, legislative support, and market ebbs and flows as we delve into the complexities of CCUS integration. The complex arena of policy interventions is set to highlight approaches that turn the aforementioned task of making CCUS economically viable into reality. More than a theoretical discourse, this study serves as a guidepost for politicians, industrial stakeholders, and environmental campaigners, providing a model for balancing economic growth and ecological stewardship.

Our research also attempts to contextualize CCUS integration within the larger context of the aforementioned environmental regulations and industry adjustments. Here, CCUS adoption becomes a cog in the machinery of efforts to find an equilibrium between economic expansion and ecological protection. We seek to unlock the benefits of incorporating CCUS while understanding the difficulties for the industry in doing so by comprehensively exploring feasibility analyses, performance evaluations, and potential hurdles. In essence, our research sets out to untangle the complex interplay of elements that determine whether CCUS will be affordable. We stand at a crossroads of environmental stewardship and industrial progress, united by a shared vision of a greener tomorrow.

The remainder of this chapter is organized as follows. Section 2 reviews the literature on climate-related disclosure and CCUS feasibility in select industries. Section 3 presents our findings and policy implications.

2. Literature Review

2.1. *Climate-related disclosures*

Following the Global Reporting Initiative (GRI) methodology, the Global Competitiveness Initiative determined that 78 percent of the top 258 global firms commenced sustainability initiatives in 2019 (Global Reporting Initiative, 2021). Figure 1 shows that international sustainability reporting rates have steadily increased over the past several years, providing information regarding the social, ecological, and economic impacts of businesses (Global Reporting Initiative, 2019).

Various conceptual approaches explain voluntary social and environmental reporting. According to the stewardship theory, businesses use standards of practice to justify their actions and inform partners of their economic, social, and environmental successes. Transparency can be used as a green energy tactic because companies with poor sustainability have greater incentives to reveal more to justify their business practices, as alluded to above. Conversely, the disclosure hypothesis postulates that companies that perform well in environmental terms are motivated to disclose their successes to the market and other constituents (Koenker and Bassett, 2007), while businesses doing poorly in this regard tend to keep quiet, so shareholders will not view them as subpar.

Various guidelines, regulations, and norms are available to organizations seeking to publish their ecological performance. The Sustainability Reports Initiative combines achievement disclosures for the economy, the

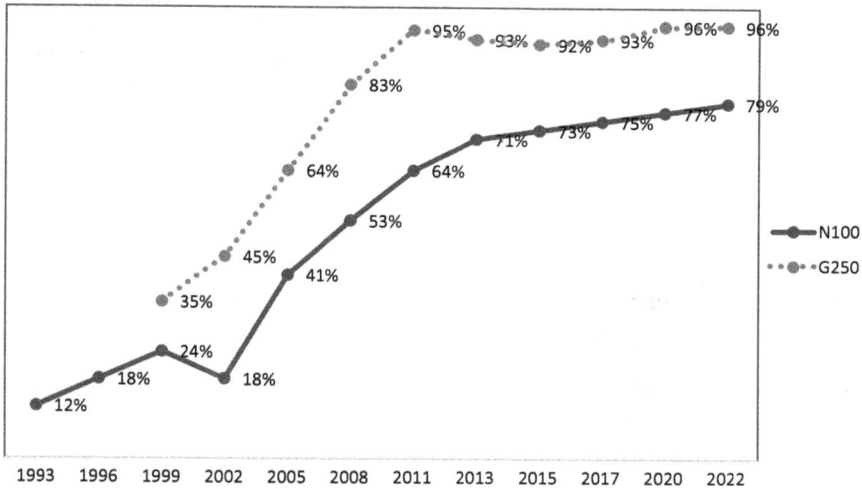

Figure 1.   Global sustainability reporting rates, 1993–2022.

Source: KPMG Survey of Sustainability Reporting 2022. KPMG International, September 2022 of 5800 N100 companies and 250 G250 companies.

environment, and society. Releases, along with substances, fuel, freshwater, industrial effluent, and wildlife, are some of the environmental protection concerns addressed (Global Reporting Initiative, 2019). The Climate Reporting Project's global climate survey emphasizes pollution and managing global warming risk, in addition to more focused issues, including carbon pollution, forestry, and irrigation programs (CDP Worldwide, 2019). A more precise discovery process that concentrates on greenhouse gases (GHGs), initiatives, and collection is the Atmospheric Carbon Protocol (World Resources Institute and World Business Council for Sustainable Development, 2005). Businesses that emit 50,000 t or more of methane gas equivalent or more (Afsharian *et al.*, 2019) are obligated to report their GHG reductions (Kao, 1999), and the Canadian province of Alberta further mandated disclosure of specific emissions of GHGs and other pollutants. Different firms have reacted differently to such regulations. Some focus on emissions plans, while others highlight environmental policy. Sustainable development can be employed to apply green initiatives in decision-making and strategic orientation and boost performance. Diverse metrics of environmental efforts may indicate sustainability stewardship in risk disclosure (Li *et al.*, 2023). Multiple aspects of businesses' green initiatives are thus communicated through the aforementioned voluntary sustainability reporting (Cade and Noon, 2003). Our initial goal is to provide a structure that allows examining ecological and carbon plans within corporate sustainability. Studying how these tactics are integrated is our third goal.

2.2. *Environmental and carbon strategies as bases for disclosure analysis*

Actions made to reduce the carbon footprint of businesses, whether freely or in compliance with law, are called sustainability practices. They are changing in light of efforts to pursue sustainable growth and limit ecological impact. While some choose reactive sustainability initiatives, others choose imaginative and proactive approaches.

Reactionary environmental initiatives concentrate on finding solutions to issues as they develop using established techniques, such as edge remedies, to reduce risk, liability, and expense (Ahmad *et al.*, 2019). Accordingly, such protections don't increase resource utilization. They merely reduce environmental costs. Edge solutions are designed to assess, treat, and reduce the environmental effect resulting from a company's operations, not to alter its methods (Lamichhane *et al.*, 2021). Such tactics thus ensure that environmental standards and standard industry practices are followed.

Green innovation strategies use contemporary techniques incorporated into management solutions to avoid ecological issues by addressing their causes (Stafford-Smith *et al.*, 2017). Businesses that use proactive environmental strategies operate quickly and creatively to obtain an advantage. Pollution can be prevented using technical innovation, such as environmentally innovative products (Aftab *et al.*, 2022). As technology has advanced significantly in recent times, proactive businesses are developing and utilizing such solutions.

The literature has also described coal strategy as a component of the sustainability strategy. This approach is envisioned as a response to worries about global warming that have galvanized authorities, the public, other groups, and businesses to consider potential means to reduce GHGs. Cost and carbon plans are accordingly being developed in response to such rising demand, including climate legislation, owners with more social responsibility, creditors, and clients. These initiatives are business solutions to environmental problems, including waste, pollutants, effluents, land use, and wildlife. By contrast, emissions plans focus on combating climate change by reducing GHGs.

A climate plan is "a complicated collection of steps to decrease the impact of an agency's commercial activities on global warming and to obtain a comparative edge over time," as stated by Sebestyén and Abonyi (2021). According to earlier studies, businesses take multiple activities to implement their objectives (Dinda *et al.*, 2000).

2.3. *CCUS utilization by various industries*

As indicated in several worldwide modeling studies, employing CCUS is an opportunity to reduce carbon emissions and achieve green management performance across a variety of industries. CCUS holds threshold, commercial, and option values in time-bound, economy-wide transitions to net-zero emissions for diverse industries (Greig and Uden, 2021).

CCUS has been assessed for market potential and barriers to deployment in the energy industry, where it is intended to mitigate CO_2 emissions from fossil fuel power plants. For instance, the Petra Nova project, implemented by NRG Energy and J.X. Nippon Oil & Gas Exploration, captures and stores CO_2 emissions from coal-fired power plants (Miyamoto *et al.*, 2017). Similarly, CCUS has been applied in the oil and gas industry for enhanced oil recovery (EOR) projects such as the Weyburn-Midale CO_2 Monitoring and Storage Project, which effectively stores CO_2 emissions in a depleted oilfield (Zaluski *et al.*, 2016). Capturing CO_2 from industrial flue gases is also crucial to mitigating GHGs and promoting sustainable practices. First, it is an essential processing step for numerous facilities such as natural gas processing plants, petroleum refineries, chemical plants, and power plants. Hydrogen plants, synthetic gas plants, coal gasification units, and tar sands generate substantial CO_2 emissions, the capture of which is vital to reducing GHGs. Second, the captured CO_2 can be utilized directly as a flooding agent for the abovementioned EOR, which is critical for sustaining current crude oil levels and ensuring steady global energy supply. Injecting captured CO_2 into oil reservoirs enhances oil recovery and extends the lifespan of existing oil fields while simultaneously permanently sequestering CO_2 underground (Tontiwachwuthikul *et al.*, 2017).

The cement industry, known for its own substantial carbon foot-print, has also recognized the need to adopt CCUS to reduce its GHGs. Doing so may be case-dependent, however, as cement plants emit more CO_2 than can be utilized. Research has focused on integrating CCUS in cement manufacturing, exploring methods for capturing CO_2 emissions from cement plants, and utilizing these emissions in various applications (Monteiro and Roussanaly, 2022). The steel industry, another significant contributor to global emissions, has embarked on initiatives such as COURSE50, FINEX-CEM, or HYBRIT, which aims to achieve carbon-free steelmaking by hydrogen reduction of iron ore (Zhang et al., 2021). The chemical industry, still another significant emitter, has also explored using CCUS to capture and utilize CO_2 emissions in various chemical processes, where it offers the potential to convert CO_2 into valuable chemicals and products, reducing emissions and promoting resource efficiency (Zhang et al., 2021). As mentioned above, the pulp and paper industry too has recognized the benefits of CCUS in mitigating its environmental impact. Research has focused on using CCUS in pulp and paper mills, contributing to sustainability (Karlsson et al., 2021).

Beyond these traditional industries, CCUS has also found applications in other sectors. The mining industry, which faces similar challenges in reducing its carbon emissions, has accordingly explored using CCUS to capture and store GHGs as well, including those associated with coal extraction (Liu et al., 2023). In the agriculture and forestry sectors, opportunities exist for carbon capture and storage in agricultural practices and biomass utilization for energy production, contributing to sustainable land use and environmental protection (Roy et al., 2023). The waste management industry can utilize CCUS for treatment of residual waste not suitable for recycling to ensure hygienic conditions, the destruction of organic pollutants, iron and aluminum recovery, and safe concentration of inorganic pollutants into air pollution control residues (Christensen and Bisinellam, 2021).

By harnessing CCUS potential and addressing its challenges, industry is able to achieve both environmental and economic benefits, aligning with the goals of a carbon-free environment and sustainable development. Policymakers, industry stakeholders, and researchers must collaborate in further exploring its potential in other sectors, addressing challenges such as cost-effectiveness, infrastructure requirements, and public acceptance (Hasan et al., 2022).

2.4. *Making CCUS economical*

CCUS' economic viability is the key to widespread adoption and successful integration into global efforts to mitigate climate change. As previously discussed, CCUS holds significant promise for reducing GHGs by capturing CO_2 from industry and electricity generation, thus preventing its release into the environment. Conversely, the economic feasibility of CCUS projects primarily relies on their financial sustainability, which entails minimizing costs, optimizing efficiency, and providing decent ROI. This section goes into numerous approaches and strategies presented in the literature to improve the economic feasibility of CCUS initiatives.

2.4.1. *Innovations and efficiency enhancements*

Innovations and efficiency enhancements are essential options for making CCUS more cost-effective. Researchers have emphasized increased CCUS use to reduce energy demand, as well as operational and overall project costs alike (Allahyarzadeh-Bidgoli and Yanagihara, 2023). Novel solvents, improved materials, and optimized capture methods show promise in increasing capture efficiency while lowering additional energy usage. Integrating CCUS with such industrial processes as the aforementioned EOR and mineralization provides opportunities to generate revenue streams that may recoup initial investments.

2.4.2. *Policy incentives and carbon pricing mechanisms*

The importance of carbon pricing mechanisms and other regulatory incentives in affecting the economics of CCUS projects cannot be emphasized enough. Governments and international organizations have begun implementing the former, which assigns a monetary value to carbon emissions, generating financial incentives for industry to minimize its carbon footprint (Zhang *et al.*, 2021). Effective carbon pricing can also create a revenue stream that offsets CCUS deployment costs and makes it more competitive than unabated emissions. Targeted subsidies, tax credits, and research grants can also aid in creating CCUS initiatives, and establishing an inviting investment climate while achieving economic sustainability (Cabrera *et al.*, 2022).

2.4.3. *Economies of scale and collaboration*

Economies of scale and coordination among industry and stakeholders can also help make CCUS more affordable. Shared infrastructure and operational efficiency typically large-scale CCUS operations to benefit from lower costs per unit of harvested CO_2 (Bae *et al.*, 2020). Collaboration can also lead to synergies in which captured CO_2 is used in such value-added applications as carbonated beverages, building materials, and chemical feedstocks. Such circular economy initiatives cut emissions and provide revenue streams, enhancing CCUS's economic viability.

2.5. *Case histories*

As sustainable development becomes more imperative, the role of CCUS in fostering a carbon-free industry becomes more critical. This section explains how several policy tools prove that CCUS is feasible, emphasizing its critical role in facilitating the transition to a greener economy.

2.5.1. *Subsidies and other incentives*

As mentioned above, subsidies and other incentives have emerged as key drivers in making CCUS economically viable, as typified by the Norwegian Longship Project. This is the world's first cross-border, open-source CO_2 transport and storage infrastructure network, allowing businesses across Europe to store CO_2 safely and permanently. The first phase will come online in 2024, with an annual storage capacity of up to 1.5 $MtCO_2$. As part of this initiative, the Norwegian government has provided subsidies and other incentives totaling USD 1.8 million to industries actively involved in CCUS (Helgesen *et al.*, 2021), removing financial obstacles to businesses seeking to integrate CCUS by reducing initial capital investment constraints. This method efficiently combines economic incentives with environmental stewardship, making a compelling business case for CCUS adoption.

2.5.2. *Tax credits and carbon pricing mechanisms*

The US 45Q tax credit, largely considered the world's most progressive CCS-specific incentive, is an example of how tax credits improve CCUS's economic viability (Victor and Nichols, 2022). This policy approach provides direct cash incentives for CCUS. The aforementioned global carbon pricing schemes also demonstrate how economic and environmental factors can converge. As mentioned above, they motivate businesses to decrease their carbon footprint by assigning a monetary value to carbon emissions. Countries including Canada have adopted it to make CCUS more viable (MacNab *et al.*, 2017).

2.5.3. *Research and collaboration*

Research and collaboration are integral to establishing CCUS as a viable climate solution, as they drive technological advancements, reduce costs, and enable commercial scaling. Research informs effective policies, addresses environmental concerns, and fosters public acceptance. Collaboration ensures holistic solutions, while global cooperation encourages widespread CCUS implementation, which is crucial to mitigating climate change.

For example, the UK's steadfast commitment to research is evident in its Clean Growth Strategy, which allocates resources for CCUS R&D, with a keen focus on bolstering technological maturity and mitigating associated risks (Geels and Gregory, 2023). It encompasses diverse sectors in such initiatives as the deployment roadmap for CCUS, an infusion of GBP 315 million into the Industrial Energy Transformation Fund, the pioneering Offshore Wind Sector Deal, establishing the Future Homes Standard that mandates the construction of forward-looking, energy-efficient new homes, and the strategic Road to Zero framework aimed at expediting electric vehicle adoption, all of which has yielded remarkable results (UKGov, 2023). Another example is Canada's 25-Year Environment Plan, which highlights the synergy between industry and government in financing via the Quest Project (Roc *et al.*, 2017). These collaborations foster an atmosphere conducive to economically sustainable CCUS deployment by sharing financial responsibility and collectively addressing technological problems.

2.5.4. *Regulatory frameworks and certifications*

Regulatory frameworks and certifications are also pivotal in making CCUS more economically viable. Clear regulations provide a predictable environment for investment by defining operational parameters, liability, and permitting processes, while also ensuring compliance, safeguarding the environment, and promoting public trust. These elements create a favorable climate for CCUS projects by encouraging financial support and enabling integration into the broader economy. For example, the aforementioned Norwegian Longship Project has one such regulatory framework (Zhang *et al.*, 2022), under which it will connect two separate CO_2 capture plants in Norway with the Northern Lights storage facility, beneath the North Sea. Northern Lights will also be able to receive CO_2 captured in neighboring European countries, helping to meet both Norway's own ambitious climate goals as well as those of the entire region. Certifications and standards such as the verified carbon standard (VCS) are also critical, for example, in ensuring that project emissions reductions are appropriately validated as part of meeting demanding environmental criteria, thereby boosting investor trust (Foster *et al.*, 2017).

2.5.5. *International collaboration and agreements*

International cooperation, typified by such initiatives as the Joint Crediting Mechanism (JCM), also helps make CCUS more economically feasible. Intended to facilitate the goal of reducing GHGs by 46 percent from FY2013 by FY2030, an ambitious target aligned with the goal of net zero by 2050, the JCM is exemplified by Japan's collaborative infrastructure development initiative. As of April 2023, 234 projects across 26 partner countries have been announced, with 138 projects online. Ajinomoto Co., Inc., Japan Petroleum Exploration Co., Ltd., JAPAN NUS Co., Ltd., and the Japan Institute of Energy Economics are all partners (Mitsubishi Research Institute, Inc., 2020). Sharing resources and costs through international collaboration leads to increased economic attractiveness, making CCUS more globally competitive.

3. Conclusions and Policy Recommendations

CCUS is critical to harmonious economic expansion and environmental protection in terms of a sustainable future. This research has accordingly thoroughly examined the economic feasibility of CCUS, including potential, problems, and policy interventions. It has become clear that symbiosis among policy, technology, and markets is the key to making CCUS viable. To this end, we offer the following recommendations:

Integrated frameworks: Policy frameworks should encompass a variety of initiatives that work together to improve the economic feasibility of CCUS. The government should combine financial, legal, and technical developmental approaches, ensuring that the industry is motivated by financial incentives and clear instructions, and equipped with the latest breakthroughs.

Carbon pricing mechanisms: Carbon pricing mechanisms, such as carbon taxes or emissions trading systems, create a market-driven incentive structure that assigns a monetary value to carbon emissions. The government should thus consider establishing or enhancing such systems to internalize the environmental costs of carbon emissions. The industry will thus be driven to minimize emissions or invest in CCUS to reduce their carbon obligations. This approach also generates revenue that may be reinvested in advancing CCUS solutions.

Research and development: Investing in R&D is critical for making CCUS more cost-effective. The government should accordingly fund programs to make CCUS more efficient, in conjunction with the private sector and research institutes. Doing so can speed the development of breakthroughs that make CCUS more economically viable.

International collaboration: The global nature of climate change demands international collaboration. The public and private sectors should accordingly share resources, knowledge, and costs connected with CCUS through collaborative efforts such as the aforementioned JCM. Countries can work together to establish a more favorable economic climate for CCUS deployment, benefiting from shared expertise and learning from one another's triumphs and challenges.

Regulatory certainty: Transparent regulatory frameworks are critical to instilling investor trust in CCUS. The government should accordingly create clear principles and criteria for developing, implementing, and overseeing programs.

Public awareness: The success of CCUS hinges on public perception and acceptance. The government and industry should accordingly invest in comprehensive campaigns to educate the public about the benefits, safety measures, and long-term potential of CCUS, thereby reducing skepticism and misconceptions. When the public is well informed, they are more likely to support policies and projects that promote CCUS, making it more feasible.

Industry collaboration: Businesses can maximize economies of scale by collaborating across sectors. Sharing resources and knowledge allows for overcoming common challenges associated with CCUS. Collaborations can also lead to circular economy models, where captured CO_2 is utilized for value-added applications, generating additional revenue streams.

The route to economic feasibility and green growth via CCUS is complex but attainable. Nations may lead sectors toward sustainable prosperity by embracing creative policies, utilizing technology advancements, and forging collaborative alliances. CCUS links economic progress and ecological balance, illustrating the connection between invention and environmental protection.

Acknowledgments

This chapter is a modified version of a forthcoming paper in the *Journal of Environmental Assessment Policy and Management*.

References

Afsharian, M., Ahn, H., and Harms, S. G. (2019). Performance comparison of management groups under centralised management. *European Journal of Operational Research*, *278*(3), 845–854.

Aftab, J., Abid, N., Sarwar, H., and Veneziani, M. (2022). Environmental ethics, green innovation, and sustainable performance: Exploring the role of environmental leadership and environmental strategy. *Journal of Cleaner Production*, *378*(September), 134639. https://doi.org/10.1016/j.jclepro.2022.134639.

Ahmad, M., Zhao, Z. Y., and Li, H. (2019). Revealing stylized empirical interactions among construction sector, urbanization, energy consumption, economic growth and CO_2 emissions in China. *Science of the Total Environment*, *657*, 1085–1098. https://doi.org/10.1016/j.scitotenv.2018.12.112.

Allahyarzadeh-Bidgoli, A. and Itizo Yanagihara, J. (2023.) Energy efficiency, sustainability, and operating cost optimization of an FPSO with CCUS: An innovation in CO_2 compression and injection systems. *Energy*, *267*, 126493.

Alola, A. A. (2019). The trilemma of trade, monetary and immigration policies in the United States: Accounting for environmental sustainability. *Science of the Total Environment, 658,* 260–267. https://doi.org/10.1016/j.scitotenv.2018.12.212.

Alola, A. A., Adebayo, T. S., and Onifade, S. T. (2022). Examining the dynamics of ecological footprint in China with spectral Granger causality and quantile-on-quantile approaches. *International Journal of Sustainable Development and World Ecology, 29*(3), 263–276. https://doi.org/10.1080/13504509.2021.1990158.

Bithas, K. and Latinopoulos, D. (2021). Managing tree-crops for climate mitigation. An economic evaluation trading-off carbon sequestration with market goods. *Sustainable Production and Consumption, 27,* 667–678. https://doi.org/10.1016/j.spc.2021.01.033.

Foster, B. C., Wang, D., Auld, G., and Roman Cuesta, R. M. (2017). Assessing audit impact and thoroughness of VCS forest carbon offset projects. *Environmental Science & Policy, 78,* 121–141.

Cade, B. S. and Noon, B. R. (2003). A gentle introduction to quantile regression for ecologists. *Frontiers in Ecology and the Environment, 1*(8), 412–420. https://doi.org/10.1890/1540-9295(2003)001[0412:AGITQR]2.0.CO;2.

Delacámara, G. (2008). Guía para decisores Análisis económico de externalidades ambientales. Cepal, 82. http://www.cepal.org/publicaciones/xml/7/33787/LCW-200.pdf.

Dinda, S., Coondoo, D., and Pal, M. (2000). Air quality and economic growth: An empirical study. *Ecological Economics, 34*(3), 409–423. https://doi.org/10.1016/S0921-8009(00)00179-8.

Douglas, P. H. and Bye, R. T. (1925). Principles of economics. *Social Forces, 4*(1), 209. https://doi.org/10.2307/3004414.

Dziejarski, B., Krzyżyńska, R., and Andersson, K. (2023). Current status of carbon capture, utilization, and storage technologies in the global economy: A survey of technical assessment. *Fuel, 342,* 127776. https://doi.org/10.1016/j.fuel.2023.127776.

Emblemsvåg, J. (2022). Wind energy is not sustainable when balanced by fossil energy. *Applied Energy, 305*(June 2021), 117748. https://doi.org/10.1016/j.apenergy.2021.117748.

Forson, J. A., Buracom, P., Chen, G., and Baah-Ennumh, T. Y. (2017). Genuine wealth per capita as a measure of sustainability and the negative impact of corruption on sustainable growth in Sub-Sahara Africa. *South African Journal of Economics, 85*(2), 178–195. https://doi.org/10.1111/saje.12152.

Geels, F. W. and Gregory, J. (2023.) Low-carbon reorientation in a declining industry? A longitudinal analysis of coevolving contexts and company strategies in the U.K. steel industry. *Energy Research & Social Science, 96,* 102953. https://doi.org/10.1016/j.erss.2023.102953.

Giorgio, C., Dickson, A., Nimubona, A.-D., Quigley, J. (2022.) Carbon capture, utilisation and storage: Incentives, effects and policy. *International Journal of Greenhouse Gas Control, 120*, 1750–5836. https://doi.org/10.1016/j.ijggc.2022.103756.

Greig C. and Uden S. (2021). The value of CCUS in transitions to net-zero emissions. *The Electricity Journal, 34*(7), 107004. https://doi.org/10.1016/j.tej.2021.107004.

Helgesen, L. I., Cauchois, G., Nissen-Lie, T., Prestholdt, M., Carpenter, M., and Røsjorde, A. (2021). CO_2 Footprint of the Norwegian Longship Project. Proceedings of the 15th Greenhouse Gas Control Technologies Conference. http://dx.doi.org/10.2139/ssrn.3821398.

Junhee, B., Yanghon, C., Jaewook, L., and Hangyeol, S. (2020.) Knowledge spillover efficiency of carbon capture, utilization, and storage technology: A comparison among countries. *Journal of Cleaner Production, 246*, 119003. https://doi.org/10.1016/j.jclepro.2019.119003.

Kai, Z., Hon Chung, L., and Zhangxin, C. (2022). Extension of CO_2 storage life in the Sleipner CCS project by reservoir pressure management. *Journal of Natural Gas Science and Engineering, 108*, 104814. https://doi.org/10.1016/j.jngse.2022.104814.

Karlsson, S., Eriksson, A., Normann, F., and Johnsson, F. (2021). CCS in the Pulp and Paper Industry — Implications on Regional Biomass Supply. *Proceedings of the 15th Greenhouse Gas Control Technologies Conference.* http://dx.doi.org/10.2139/ssrn.3820355.

Kao, C. (1999). Spurious regression and residual-based tests for cointegration in panel data. *Journal of Econometrics, 90*(1), 1–44. https://doi.org/10.1016/S0304-4076(98)00023-2.

Koenker, R. and Bassett, G. (2007). Reqression quantiles. *Jstor, 46*(1), 33–50.

KPMG International. (2022). Big shifts, small steps — KPMG. Survey of Sustainability Reporting. assets.kpmg.com/content/dam/kpmg/se/pdf/komm/2022/Global-Survey-of-Sustainability-Reporting-2022.pdf.

Onarheim, K., Santos, S., Kangas, P., Hankalin, V. (2017). Performance and cost of CCS in the pulp and paper industry part 2: Economic feasibility of amine-based post-combustion CO_2 capture. *International Journal of Greenhouse Gas Control, 66*, 60–75. https://doi.org/10.1016/j.ijggc.2017.09.010.

Lamichhane, S., Eğilmez, G., Gedik, R., Bhutta, M. K. S., and Erenay, B. (2021). Benchmarking OECD countries' sustainable development performance: A goal-specific principal component analysis approach. *Journal of Cleaner Production, 287*. https://doi.org/10.1016/j.jclepro.2020.125040.

Li, G., Jing, Z., Feng, Y., and Li, J. (2023). Drivers of risk correlation among financial institutions: A study based on a textual risk disclosure perspective. *Economic Modelling,* 106468, https://doi.org/10.1016/j.econmod.2023.106468.

Liu, Y. and Feng, C. (2023). Promoting renewable energy through national energy legislation. *Energy Economics*, *118*(January), 106504. https://doi.org/10. 1016/j.eneco.2023.106504.

Rock, L., O'Brien, S., Tessarolo, S., Duer, J., Bacci, Vicente, O., Hirst, B., Randell, D., Helmy, Md., Blackmore, J., Duong, C., Halladay, A., Smith, N., Dixit, T., Kassam, S., Yaychuk, M. (2017.) The Quest CCS Project: 1st Year Review Post Start of Injection. *Energy Procedia*, *114*, 5320–5328. https://doi.org/10.1016/j.egypro.2017.03.1654.

MacNab, J., Flanagan, E., Kniewasser, M., and Hastings-Simon, S. (2017). Putting a price on carbon pollution across Canada: Taking stock of progress, challenges, and opportunities as Canada prepares its national carbon pricing benchmark. Pembina Institute. http://www.jstor.org/stable/resrep02881.

Mitsubishi Research Institute, Inc. (2020). FY2019 Study on the Infrastructure Development Project for Acquisition of JCM Credits (International Cooperation in CCUS) Report, www.meti.go.jp/meti_lib/report/2019FY/000272.pdf.

Miyamoto, Cole, M., Tatsuya, T., Masayuki, I., Takuya, H., Hiroshi, T., Takahito, Y., and Takashi, K. (2017). KM CDR Process TM Project Update and the New Novel Solvent Development. *Energy Procedia*, 114, 5616–5623. https://doi.org/10.1016/j.egypro.2017.03.1700.

Monteiro J., and Roussanaly S. (2022). CCUS scenarios for the cement industry: Is CO_2 utilization feasible? *Journal of CO_2 Utilization*, *61*, 102015. https://doi.org/10.1016/j.jcou.2022.102015.

Faruque Hasan, M. M., Zantye, Manali S., and Monzure-Khoda, K. (2022). Challenges and opportunities in carbon capture, utilization and storage: A process systems engineering perspective. *Computers & Chemical Engineering*, *166*, 107925. https://doi.org/10.1016/j.compchemeng.2022.107925.

Nadejda, V. and Nichols, C. (2022). CCUS deployment under the U.S. 45Q tax credit and adaptation by other North American governments: MARKAL modeling results. *Computers & Industrial Engineering*, *169*, 108269. https://doi.org/10.1016/j.cie.2022.108269.

Ofori, E. K., Onifade, S. T., Ali, E. B., Alola, A. A., and Zhang, J. (2023). Achieving carbon neutrality in post COP26 in BRICS, MINT, and G7 economies: The role of financial development and governance indicators. *Journal of Cleaner Production*, *387*(January), 135853.

Roy, P., Mohanty, A., and Misra, M. (2023). Prospects of carbon capture, utilization and storage for mitigating climate change. *Environmental Science: Advances*, *2*, 409–423.

Sebestyén, V. and Abonyi, J. (2021). Data-driven comparative analysis of national adaptation pathways for sustainable development goals. *Journal of*

Cleaner Production, 319(December 2020), 128657. https://doi.org/10.1016/j.jclepro.2021.128657.

Shahbaz, M., Nasir, M. A., and Roubaud, D. (2018). Environmental degradation in France: The effects of FDI, financial development, and energy innovations. *Energy Economics, 74*, 843–857. https://doi.org/10.1016/j.eneco.2018.07.020.

Shiqi, L., Tong, L., Sijian, Z., Ran, W., and Shuxun, S. (2023). Evaluation of carbon dioxide geological sequestration potential in coal mining area. *International Journal of Greenhouse Gas Control, 122*, 103814. https://doi.org/10.1016/j.ijggc.2022.103814.

Stafford-Smith, M., Griggs, D., Gaffney, O., Ullah, F., Reyers, B., Kanie, N., Stigson, B., Shrivastava, P., Leach, M., and O'Connell, D. (2017). Integration: the key to implementing the Sustainable Development Goals. *Sustainability Science, 12*(6), 911–919. https://doi.org/10.1007/s11625-016-0383-3.

Svensson, E., Wiertzema, H., and Harvey, S. (2021). Potential for negative emissions by carbon capture and storage from a novel electric plasma calcination process for pulp and paper mills. *Frontiers in Climate, 3*, 705032. https://doi.org/10.3389/fclim.2021.705032.

Thomas H. C. and Bisinella, V. (2021). Climate change impacts of introducing carbon capture and utilisation (CCU) in waste incineration. *Waste Management, 126*, 754–770. https://doi.org/10.1016/j.wasman.2021.03.046.

Tontiwachwuthikul, P. T., Zeng, F., and Chan, C. W. (2017). Special issue on, Carbon Capture, Utilization and Storage (CCUS): Technological developments and future opportunities for petroleum industry. *Petroleum, 3*(1), 1–2. https://doi.org/10.1016/j.petlm.2017.02.002.

UK. Gov (2023). https://www.gov.uk/government/publications/clean-growth-strategy.

Xinyu, Z., Kexin, J., Jianliang, Z., and Ziyu, G. (2021). A review on low carbon emissions projects of steel industry in the world. *Journal of Cleaner Production, 306*, 127259. https://doi.org/10.1016/j.jclepro.2021.127259.

Zaluski, W., El-Kaseeh, G., Lee, S.-Y., Piercey, M., and Duguid, A. (2016). Monitoring technology ranking methodology for CO_2-EOR sites using the Weyburn-Midale Field as a case study. *International Journal of Greenhouse Gas Control, 54*. 10.1016/j.ijggc.2016.06.012.

Zhang, W., Dai, C., Luo, X., and Ou, X. (2021). Policy incentives in carbon capture utilization and storage (CCUS) investment based on real options analysis. *Clean Technologies and Environmental Policy, 23*. 10.1007/s10098-021-02025-y.

Chapter 11

The Political Economy of Carbon Capture, Use, and Storage

David Havyatt*, Rabindra Nepal†, and David Johnstone‡

School of Business, Faculty of Business and Law, University of Wollongong

*dsh235@uowmail.edu.au

†nepal@uow.edu.au

‡djohnsto@uow.edu.au

Abstract

Policymakers often overlook the political and economic challenges pertaining to the development and deployment of carbon capture, use, and storage (CCUS), even as they consider CCUS a technological solution to combat climate change. This chapter accordingly draws on a political economy analysis to question this notion, synthesizing five important propositions. We identify the CCUS analysis frames of technology, markets, environment, and society. Two important policy lessons are drawn regarding CCUS best practices: avoiding strategy lock-ins and actively working on alternate energy sources. The chapter concludes that political economy challenges inherent to CCUS may mean that overt reliance on its eventual development may be a high-risk strategy where climate change is concerned. Conversely, the policy goal of achieving a secure energy supply, especially in developing economies, may be well served as CCUS will prolong continued cheaper coal usage instead of abandonment.

Keywords: CCUS, climate change, energy security, best practices.

271

1. Introduction

Carbon capture, use, and storage (CCUS) has passionate supporters and opponents alike, as well as many skeptics. It is inherently political. This suggests a political economy lens is appropriate to analyze it. Some three-quarters of global greenhouse gas emissions (GHGs) come from generating energy, i.e., power plants, as well as direct combustion heat and transport. The emissions reduction challenge is primarily a story of energy systems, making climate and energy policies key (Scrase *et al.*, 2009). Many OECD economies, including Australia, have therefore aligned their climate and energy policies in pursuit of net-zero targets. The Australian government legislated Australia's emissions reduction target of 43 percent and net-zero emissions by 2050 in 2022.

Observing that "energy markets are a perfect topic for a comparative law and economics approach," Bellantuono (2016) notes that two waves of reforms have deeply transformed energy systems in recent decades. The first, generally called liberalization, saw energy systems restructured and competition introduced in various forms. The second is the reforms aimed[1] at making energy systems more environmentally sustainable. However, even though the first of these has gone global, under what has been called "textbook reform architecture" (Joskow, 2006; Littlechild, 2006) and the "standard reform model" (Kessides, 2004), there are significant differences between them. Moreover, despite accolades for these policies, the pace of reform has not been uniform, nor have the outcomes always been positive (Williams and Ghanadan, 2006), which Nepal and Jamasb (2012) note is highly interdependent with other reforms. In brief, institutions matter.

---

[1] Labeling these first and second is not intended to imply that one necessarily came before the other. In Australia, the meeting of First Ministers of the Commonwealth, States, and Territories ("the Special Premiers Conference") in October 1990 took the first steps in electricity reform as well as adopted first principles on responding to the greenhouse effect. On the latter, the Communique of the meeting noted that "The States strongly supported Australian efforts to get an international convention on climate change, and they agreed with the Commonwealth that Australia should not implement response measures that would have net adverse economic impacts nationally or on Australia's trade competitiveness, in the absence of similar action by major greenhouse gas producing countries" (Havyatt, 2022).

Glachant and Perez (2011) identified seven differences in implementation, three of which Bellantuono focused on. These are the definition of the boundaries between monopolistic and competitive activities, the allocation of regulatory tasks among decision-making levels, and the sequencing of the reforms. These differences arise from constraints that determine the solution space, including physical, economic, and institutional factors, and an institutional context that includes related policies and restructuring objectives, as shown in Figure 1 (Correljé and De Vries, 2008). All of these describe the approach to the second energy reform.

Technological solutions such as CCUS can play a key role in decarbonization and therefore addressing the challenge of global climate change. However, the development and deployment of climate-friendly technology largely depend on deeper and often complicated political economy contexts which this chapter aims to unravel. The remainder of this chapter is structured as follows. Section 2 discusses the political economy of climate change and energy as a pathway to analyzing the political economy of CCUS and accordingly proposes a frame analysis to study CCUS. Section 3 discusses four identifiable frames for said CCUS analysis. Section 4 lays out five propositions on the political economy of CCUS. Section 5 concludes the chapter.

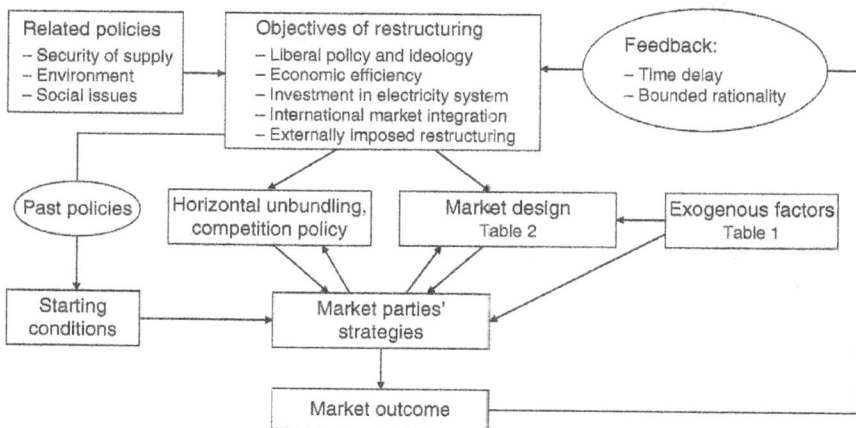

Figure 1. Institutional context of electricity market reform.

Source: Correljé and De Vries (2008).

2. Political Economy of Climate Change

Before we analyze the political economy (PE) of CCUS, we must develop a framework for the PE of energy and climate change. Section 2 is divided into the following subsections to this end.

2.1. *Political economy and political economics*

PE was, of course, the term by which the discipline of economics was known until the late 19th century. The change in nomenclature came with the marginalist revolution and the treatment of economics as a science, argued by some as having been driven by recent developments in physics (Mirowski, 1991; Weintraub, 2002), a situation some came to describe as "physics envy" (Schabas, 1993).

As Frieden (2020) notes, "By the 1970s, however, it was clear that the separation between the economic and political spheres was misleading." This resulted in a restoration of PE under the description, "Political economy is about how politics affects the economy and the economy affects politics."

"Political economy analysis is concerned with the interaction of political and economic processes within a society: the distribution of power and wealth between different groups and individuals, and the processes that create, sustain and transform these relationships over time" (Collinson, 2003).

The Australian Department of Foreign Affairs and Trade, in a guidance note on political economy analysis (DFAT, 2016), states:

> Political economy analysis is about understanding the political dimensions of any context and actively using this information to inform policy and programming. Politics is the formal and informal ways through which contestation or cooperation occurs in a society. Political processes are dynamic and occur at all levels of society. Political economy analysis involves looking at the dynamic interaction between structures, institutions and actors (stakeholders), to understand how decisions are made.

The separation of economics from political and other social analysis arises from the methodological assumptions underpinning neoclassical, i.e., orthodox, theory. These are methodological individualism, methodological instrumentalism, i.e., maximizing preference satisfaction, and methodological equilibration (Arnsperger and Varoufakis, 2006). However, the orthodox "theoretical model of market exchange under competitive conditions fails to illuminate the world in which we actually live" (Stilwell, 2012, p. 3).

These failures of neoclassical economics are well-documented (Keen, 2011; Lawson, 2013, 2017; Quiggin, 2010), and even so well-recognized that microeconomic theorists now eschew Friedman's (1953) claim for economics to be a positive science and regard their models as normative, i.e., how markets would behave if the assumptions could be made to apply. A striking example of this failure of orthodoxy comes from emissions reduction policy. Consumers do not make "no regrets" energy efficiency investments that would also reduce emissions despite net present value being positive (Marechal, 2012, pp. 28–30). This "energy-efficiency gap" (Jaffe and Stavins, 1994) or "efficiency paradox" (DeCanio, 1998) suggests a need for more comprehensive theories of consumer investment behavior. The paradox can be attributed to "market failures," including misplaced incentives, imperfect information, and lack of access to capital (Brown, 2001; Jaffe and Stavins, 1994), as well as the "bounded rationality" (Simon, 1997 (1945)) of businesses or "heuristics" used by consumers (Tversky and Kahneman, 1974). It is also important to note the market failure approach to intervention is generally poorly founded because all markets suffer to some degree from deviations from the assumption of the "free market" model (Kay, 2007), and all markets are subject to some design rules.

The consequence of these observations is that "political economy" is not studied by simply adding a political analysis to economic orthodoxy. A different kind of economics is also required, one that embeds the political and social within the theory. These alternative theories are usually taught or written about under the headings of "political economics" (Heilbroner, 1970), "heterodox economics" (Lawson, 2006), or "pluralist economics" (Fullbrook, 2008; Harvey, 2020). The last of these terms emphasizes the possible use of multiple alternate theories, which is not as surprising as it may seem. Physics has quantum, classical, or Newtonian, and relativistic theories of mechanics, all of which are used by physicists in particular "domains" where certain characteristics apply; e.g., classical mechanics is still fine for everyday objects and even our solar system. Additionally, while political economics can also be framed as interrogating the extent to which government is or should be in the economy (Alt and Chrystal, 1983), not all answers outside the neoclassical view argue for more government involvement. Notably, Austrian economics, exponents of which played a major role in market reforms in the UK, argues for less (Littlechild, 1979).

These alternatives are systems that do not adopt at least one of the axioms of the orthodox model. Austrian economics focuses on the economy as a perpetually dynamic system never in equilibrium, institutional economics violates the principles of individualism, and behavioral economics looks to aspects other than utility maximization that influence consumer and producer decision-making. Evolutionary economics, most broadly described, violates all of these.

The descriptions of political economy at the start of this subsection can also be understood in terms of their violation of these axioms of orthodox economics. Institutions, as understood by the American institutionalists, include the unstated rules and habits of markets that make the assumption of methodological individualism invalid. Similarly, the focus of Austrian economists on the dynamic nature of markets is substitute for the invalid assumption of methodological equilibration. Finally, the presence of network effects and scale economies invalidates the assumption of methodological instrumentalism. Every decision made by every producer and consumer changes the circumstances for all other individuals, including decisions involving some degree of economic or political power. The combination of these three is captured partially by evolutionary approaches. The tools provided by these approaches can only be a partial analysis, however, as none has been developed into as complete a theory as the neoclassical juggernaut.

2.2. *Political economy analysis for energy security policy*

Despite the overwhelming theory and evidence of climate change, it can be argued that the economics discipline has added little to the debate (Harvey, 2020, p. 148). One notable exception is economics Nobel laureate William Nordhaus, who introduced a quantitative model, popularly known as an integrated assessment model, which describes the global interplay between the economy and the climate. Nonetheless, where there is engagement, there are deep divisions between a neoclassical approach, i.e., environmental or environmental and resource economics, and a non-neoclassical and multi-disciplinary approach, i.e., ecological economics (Harvey, 2020; Van den Bergh, 2001). The neoclassical approach has been typified by calls for a "price on carbon," reflecting the simple idea that the way to manage pollution is to turn it into a property right, after Coase (1959, 1960). Responses argue that the socioeconomic context of markets includes power structures and ideologies that partially determine outcomes (Bate, 2001). Hence, initiatives such as a price on carbon place too great a reliance on market mechanisms (Stilwell, 2011).

What has been called the "under-theorization of the politics of energy in the social sciences" (Graaf *et al.*, 2016, pp. 8–9) can be essentially explained as follows. First, many energy experts are not much engaged in theory of any kind, being more concerned with short- and medium-term market reactions. The "grey literature," i.e., non-peer reviewed papers and reports, constitutes 60 percent of citations in three leading journals from 1999 to 2013 (Sovacool, 2014). Second, the social scientist must acquire much science and engineering knowledge to accompany social theory. Third, the multi-faceted nature of energy policy defeats unidimensional analysis, such as the acknowledged failure of neoclassical prescriptions. Social factors affect both the demand and supply sides. See, for example, literature that analyzes choices of energy sources for cooking (Akintan *et al.*, 2018; Bharadwaj *et al.*, 2021; Liao *et al.*, 2019; Malakar *et al.*, 2018; Masera *et al.*, 2000). Finally, the multi- or interdisciplinary study needed does not fit easily into the academic reward system (Lutzenhiser and Shove, 1999; Spreng, 2014).

The starting point for energy policy is the concept of "energy security," which the International Energy Agency (IEA) defines as "the uninterrupted availability of energy sources at an affordable price."[2] This implied that a nation, a community, a company, or an individual has sufficient energy resources available to meet their energy needs. The meaning of energy security is unclear, however. One study identified 45 different definitions (Sovacool, 2011, pp. 4–6), while Chester's (2010) examination of "explicit and inferred definitions" found the concept of energy security "inherently slippery because it is polysemic in nature, capable of holding multiple dimensions and taking on different specificities." The fuzzy nature of the concept of energy security can be further gleaned by examining varying national and international perspectives (Valentine, 2010). The latter include existing supplies, augmentation of energy trade, and finding new supplies. They can also be stated as balancing availability, accessibility, and affordability, or simply focusing on

---

[2] https://www.iea.org/topics/energy-security, retrieved 10 November 2022.

the bounds imposed by the planetary ecosystem. The former range from the purely technical, sometimes called system security, to the economic desire for risk-free access to the cheapest forms of energy. National perspectives must also consider ecological ambitions, however, including pollutants and habitat destruction over and above GHGs, as well as social considerations, including equity and the capacity of energy use to build social capital, for example, the importance of artificial light in being able to study.

While concepts of fuel availability, energy conversion, and distribution feature highly in the engineering interpretation and price features highly in the economic, there is also a geopolitical aspect to need. As noted in Subsection 2.1, there are abundant "no regrets" options to reducing energy consumption in developed economies, while the definition of energy security in developing economies encompasses greater than current consumption. As the IEA notes, there is also a time dimension to the concept. "Long-term energy security mainly deals with timely investments to supply energy in line with economic developments and environmental needs, [while] short-term energy security focuses on the ability of the energy system to react promptly to sudden changes in the supply-demand balance."[3]

To this simple analysis is added the sustainable energy security requirement, which in the current era includes climate change and peak oil (Bardi, 2009, 2019; Chapman, 2014). Environmental sustainability, affordability, and energy security form a policy trilemma. Policymakers inevitably face trade-offs when a particular energy policy goal gets prioritized. Indeed, irrespective of the need to reduce GHGs, continued reliance on fossil fuels must end at some point. The BP Statistical Review of World Energy 2022 estimates current reserves-to-product ratios for oil, gas and coal to be 54, 49, and 139 years, respectively).[4]

Having established the concept of the political economy of energy policy as the study of energy security, we now turn to the tools for the task. The DFAT structure outlined in Section 2.1 called for analyses based on structures, actors, and institutions. To these, we must now add frames, a meta-layer of analysis.

---

[3] https://www.iea.org/about/energy-security, retrieved April 9, 2023.
[4] https://www.bp.com/en/global/corporate/energy-economics/statistical-review-of-world-energy.html, retrieved November 10, 2022.

2.2.1. *Frames*

In any area of policy development, differences can emerge about proper responses. Schön and Rein (1994, pp. 3–4) distinguish between disagreements and controversies, wherein the latter is the class of policy dispute that is "immune to resolution by appeal to the facts." Schön and Rein (1994, p. 23) also propose that controversies arise when contending parties hold conflicting frames. The disputes are "resistant to resolution by appeal to facts or reasoned argumentation because the parties' conflicting frames determine what counts as a fact and what arguments are taken to be relevant and compelling." By "frame," they mean in this context "underlying structures of belief, perception, and appreciation." They note that by "appreciation," they are referring to Vickers' (1965) concept of an "appreciative system," which he used to refer to "a set of readinesses to distinguish some aspects of the situation rather than others and to classify and value these in this way rather than that." It is this sense, that the decision-maker is applying, usually unconsciously, a filter on aspects of the situation under analysis that gives the word "frame" its cogency herein. The frame refers to that part of both the problem and the evidence that the decision-maker uses.

The concept of frame that they use is tied to the idea of framing as applied in behavioral economics (Tversky and Kahneman, 1985), which finds that the "decision frame," i.e., the decision-maker's conception of the acts, outcomes, and contingencies, associated with a choice, is controlled in part by the formulation of the problem and the "norms, habits and, personal characteristics" of the decision-maker. While explaining Schön and Rein's observation that frames are usually tacit and exempt from conscious attention and reasoning, this latter construction goes further, in that the way the problem is put to the decision-maker helps create the frame.

Sovacool and Brown (2015) identify eight frames, or "epistemic cultures," used in energy policy, which they label technological optimists, free market libertarians, defenders of national security, energy philanthropists, justice advocates, neo-Marxists, and conscientious consumers. Graaf and Zelli (2016) claim these frames "bear close resemblance to four major worldviews of the global political economy of the environment" identified by Clapp and Dauvergne (2011). These four worldviews, market liberals, institutionalists, bioenvironmentalists, and social greens are broader constructs than frames.

The importance of frames, however, lies in awareness, not classification. Resolving a policy controversy can be described as a "wicked problem," an expression originating in social planning (Churchman, 1967; Rittel and Webber, 1973; Skaburskis, 2008) and which has come to refer to anything that is "difficult or impossible to solve because of incomplete, contradictory, and changing requirements that are often difficult to recognise."[5] An Australian government publication included climate change as an example (APSC, 2007). We can then understand frames as the theory, or more generally the often understated baggage, used for policy analysis. Recognition of frames provides the opportunity to use them as tools for examining problems. However, as Schön and Rein (1994, p. 29) explain, "Frames are not free-floating but are grounded in the institutions that sponsor them, and policy controversies are disputes among institutional actors who sponsor conflicting frames."

2.2.2. Structures

As described by DFAT (2016), structure in political economy analysis refers to "the more enduring specifics of the context that change slowly, such as global influences, natural resource endowment, demographic shifts, historical legacies, social-cultural factors, and technological progress."

Natural resource endowments are particularly important in energy policy. For example, the decisions to create government owned Electricity Commissions in the Australian states of Victoria and South Australia were motivated by the desire to utilize local brown coal because of the unreliability of black coal supplies from New South Wales. Similarly, demographic shifts and economic progress determine necessary energy supplies.

Whether socio-cultural factors should be classed as structures or institutions depends on how amenable to change they are. An example of one such factor that might be considered structural is cultural resistance to cooking with modern fuels in some parts of South Asia (Akintan et al., 2018; Boudewijns et al., 2022; Malakar et al., 2018). The current state of technology and the rate at which technology might change are also

---

[5] This phrase comes from the Wikipedia entry, retrieved November 10, 2022. It is also treated on other sites as an accepted definition.

structural variables, as countries or utilities cannot implement CCUS strategies with technology that does not yet exist.

2.2.3. *Institutions*

Institutions is a difficult word because it has both a narrow meaning, referring to something substantial, such as a business organization or a governance entity, and a broader meaning, referring to these structures as well as the rules that govern their operation, and the informal rules and habits of social discourse. It is the latter meaning that applies herein. Environments and governance mechanisms are two essential components of institutions (Williamson, 1995). Environment is the "rules of the game," which can be both formal or informal, while governance mechanisms refer to the ways in which business units organize, compete and cooperate.

The highest institutional level is the concept of government, which is distinct from governance structures. Government itself, the right of some authority to make rules for society, or more generally to exercise coercive state power, is a social construct of relatively recent vintage. In *The Origins of Political Order*, Fukuyama (2011) identifies three elements of political order, i.e., government, being the state, the rule of law, and accountable government. While not all of these elements exist together, in full they actually emerge around the same time as mankind started to develop energy systems.

The institution of government differs by setting, and has dimensions of scope, i.e., national, provincial, and local, and type, i.e., democratic and autocratic. Instances of government are actors whose roles are bound by the governance institution (Averchenkova and Nachmany, 2017), including government and the agencies it creates. The institution of governance is highly significant in climate policy, with one study suggesting that improving average energy governance quality 10 percent may raise energy efficiency 9.2 percent (Barrera-Santana *et al.*, 2022).

Closely related to the institution of government is the institution of societal values. These can range from the explicitly stated, i.e., "Liberty, equality, fraternity," or "that all men are created equal, that they are endowed by their Creator with certain unalienable Rights, that among these are Life, Liberty, and the pursuit of Happiness," to the often unstated but widely observed. They incorporate attitudes to equality and equity and attitudes toward the environment. The latter can be expressed as being only about intergenerational equity or as a non-human- or panpsychic-centered value set (Grusin, 2015).

The next institutional level is the framework for production and consumption of goods. While this is typically a market economy the world over, the extent of constraints on markets, and of direct government involvement in production or consumption decisions, varies. As such, the electricity market reforms that occurred in many countries in the 1980s and 1990s were major institutional reforms (Kessides, 2004; Sioshansi and Pfaffenberger, 2006).

Bridging the institutions of government and the economy is the field of inter-governmental relations and trade. Multilateral treaties are an institution, while entities such as the United Nations (UN) and the Association of South-East Asian Nations (ASEAN) are institutional actors. Similarly, the General Agreement on Tariff and Trade (GATT) is an institution, while the World Trade Organization (WTO) is an actor. Various groupings of countries according to country type are also institutions, such as the distinction between developed and developing nations.

The technologies of production and consumption are institutional variables, as are the social functions of energy use. Hence, energy can be viewed through a socio-technical institutional lens (Büscher *et al.*, 2018). The inclusion of human activities creates the concept of the "sociotechnical regime" (Geels, 2002), defined as "the semi-coherent set of rules" that govern energy systems. These rules function like paradigms and heuristics in shaping the way engineers, economists, and others concerned with policy think about systems. For example, if many machines connected to an electricity grid consist of moving parts, i.e., turbines and motors, then it is important to maintain system frequency within a tight band, to avoid unleashing forces inside those machines, making this a policy focus and an assumed requirement. This is less of an issue, however, if all generation and load is inverter-connected, as is likely to be true in future. Overall, the mix of energy sources is an institutional variable, as is the use of energy as heat, light, and power. Within the socio-technical regime there are many actors involved in energy production, distribution, and consumption.

A subset of technology and social institutions is technological change, which is also institutional in nature as the pace of innovation, adaption, and diffusion is determined by factors that are more than mere individual actors' utility-maximizing decisions. The pace or opportunity for change is also mediated by factors underpinning path dependence, including bounded rationality and economies of scale.

While the assumption in liberal democracies is that major producers will be common-stock companies, many institutional dimensions remain. The first is that these entities may be state-owned enterprises, state ownership may take other economic forms, such as government departments or the statutory commissions once popular in Australia. The second is that companies may merely be subsidiaries of global corporations or, increasingly, owned by a form of private equity, typically pension funds such as Ontario Teachers or asset investment managers such as TPG Capital or Blackrock. The third variable is the extent of vertical integration among generation, transmission, distribution, and retail. Note that the reform model typically separated these functions (Kessides, 2004).

The last of these can be summarized as finance, the last element of the institutional framework of production. While the concept of a market economy is typically also described as capitalism, we are already deep into its third phase. In the first phase, the owners of the means of production were capitalists, individuals who, either on their own or in small partnerships, built mines, factories, and railroads. Only in the middle of the 19th century did the common-stock company become anything other than a bespoke vehicle authorized by specific legislation on a per company basis. This second stage of capitalism was typified by stock exchange listings, and developed, largely in one country, into the multinational corporation that still often involved listing subsidiaries in new markets as ways to provide capital for growth. This managerial capitalism represented a high point of corporate transparency and facility for government oversight. The currently third phase of capitalism consists of large transnational enterprises with highly complex internal structures. While some remain listed on stock exchanges, others increasingly are not. Some are conglomerates ultimately held by the aforementioned state-owned enterprises, many of which are owned by democratic governments, while some are held by the aforementioned private equity funds that access personal wealth and pension funds (Kocka, 2016; Micklethwait and Wooldridge, 2005).

Associated with these changes in capitalism has comes a change in firms' institutional role. Ownership capitalism had its focus on production, where the interest was in building enterprises and intergenerational wealth. Managerial capitalism's focus moved to increasingly shorter-term returns. Finally, financial capitalism's focus is on returns available from restructuring, refinancing, and tax avoidance.

2.2.4. *Actors*

The primary, though often unacknowledged, actors in energy systems are the consumers for whom heat, light, and power are provided, and whose other goods and services are provided by individuals or combinations using heat, light, and power to make those goods. While every person within an economy is a consumer, they are also citizens. They have interests in energy systems beyond consumption, being concerned energy systems' role in prosperity, equity, and environmental quality. The secondary actors seek to represent these consumers' or citizens' interests, typically political parties, charities, and interest or advocacy groups.[6]

On the production side, the actors are collectives of individuals operating as companies, however structured or owned, which can typically be considered to consist of investors, managers, and workers, sharing some objectives while often being in conflict over others. Each subgroup has associated organizations involved in policy and advocacy, with industry associations typically representing managers and unions representing workers. Similar firms exist on the business consumption side with similarly subcategories and advocacy structures. Firms also engage with or support, and sometimes join, citizens advocacy groups. The industries that supply the producers, notably mining and technology firms, constitute a different group of actors that have the same characteristics as other firm-type actors. These are not necessarily local firms, however, and hence are associated with actors in the governance space. Government institutions include local executives, legislatures, and independent regulatory commissions and courts. They include laws enacted for the sector and the common interpretation of those laws. They also include relevant foreign governments as well as regional and global multilateral bodies tasked with general global security, energy security, and climate change.

2.2.5. *Summary*

We have now established a collection of energy policy analysis tools. In summary, the wicked problems or policy controversies must be examined through different frames, with the analysis thereof identifying constraints

---

[6]In the UK, all not-for-profit entities are considered charities. Hence, many policy advocacy groups are called charities, whereas in Australia at least, charity explicitly means organizations involved in giving aid in some form or other, including medical research.

that existing institutions and the preferences of different institutional actors impose.[7] The complexity of the political economy of climate change versus that of energy security alone is the need for effective international deals among diverse actors conducted in an environment rich with possible time inconsistency (Rothenberg, 2012).

Writing three years before the Rio Earth Summit,[8] Quesada (1989) contemplated the first of these challenges, writing, "The political economy of climate change will call into question all major issues in the global agenda, and humanity has a unique opportunity to settle old problems and design an area of international cooperation of unparalleled dimensions." It took a further 23 years to get to the first defined goal, with the Paris Agreement.

2.3. *From energy security to climate change*

The leading contending frames are obviously energy security for further economic growth, and reversing global warming. The purpose of this chapter is to use the tools to analyze CCUS, rather analyze energy security policy.

2.3.1. *The market approach to greenhouse gas emissions*

As mentioned previously, considering the political economy of climate change itself, other than as an aspect of energy security policy, introduces additional perspectives. One of these is the institutional lens that defines capitalism and its economics[9] as the core problem (Xie and Cheng, 2021). As noted above, the orthodox response to an externality is to assign it a property right. The orthodox toolkit suggests that the price that would be

---

[7]This approach is similar to that of Jakob *et al.* (2020), which used actors, actors' objectives, and context. Their use of "context" covers our definition of "institution" as covering both formal and informal rules and behaviors.

[8]United Nations Conference on Environment and Development, Rio de Janeiro, Brazil, June 3–14 1992, https://www.un.org/en/conferences/environment/rio1992, at which the United Nations Framework Convention on Climate Change (UNFCCC) was created.

[9]Under the title *Capitalism and Its Economics: A Critical History*, Dowd (2004) makes the case that orthodox economics is more an ideology in support of capitalism than a science.

applied to "buy" a unit of reduction in GHGs is accordingly determined by the sum of the discounted future benefits.

The implementation of carbon prices, or emissions trading schemes, exhibits the same degree of variety around the world as does energy reform, as outlined in the Introduction. Rudolph and Aydos (2021) identify eight characteristics in which ETS design may vary:

- Coverage: decisions about which GHGs and participants, i.e., polluters, are to be included.
- Cap: who sets it and at what level.
- Allocation: whether issued for free or at auction.
- Revenue use: to what purpose revenue raised is put, especially whether redistributed to users.
- Flexibility mechanisms: whether a polluter can bank or borrow allocations for or from a future time period.
- Price management: especially the question of price floors or ceilings.
- Compliance systems for monitoring, registration and verification.
- Linking: the extent to which two independent ETS enable participants to use permits in one market to offset liability in the other, i.e., inter-system trading.

Nevertheless, Rudolph and Aydos assert that a Sustainable Model Rule can be developed to simultaneously achieve the goals of environmental effectiveness, economic efficiency, and social and climate justice. This view, however, is based on a view of "public choice" which they describe as "the economic theory of politics." While they do propose an extension of public choice theory (p. 213) by recognizing differences within classes of economic actors, this does not sufficiently account for the complexity of decision-making, that is, the politics inherent in economic decision-making.

There are other assumptions inherent in the design of a market for "bads," rather than goods. The first is that no means exists to choose an appropriate discount rate to convert future outcomes into present-day values. Less obvious is that there is no meaningful way in which to assess future benefits, as these ultimately depend on consumer preferences and we have no way of knowing what consumers' future preferences will be.[10]

---

[10] This criticism of both choosing a social discount rate and the impossibility of knowing future consumer preferences of course invalidates any cost-benefit analysis methodology.

The heterodox critique centers on neither of these mechanics, however. Instead, it focuses on the core assumption that the "multi-dimensional nature of human well-being [is] interchangeable into utility units" (Xie and Cheng, 2021, p. 318). Market fetishism pervades the thinking in this regard, ignoring the artificiality of the worldview necessary to make neo-classical market theory work (Mallin, 2011).

Xie and Cheng (2021) identify the following challenges for emissions trading schemes. The first is that they rely on converting the climate impact of all GHGs to CO_2 equivalents, using the Intergovernmental Panel on Climate Change (IPCC) calculation of global warming potential (GWP). This task is imprecise because the effects of different gases are contestable, and the calculation is made over an arbitrary time period of 100 years (MacKenzie 2009, p. 446).[11] Second, the schemes presuppose equivalent emissions reductions at different times and locations, typically paying more as carbon budgets decline while the value in reduced warming is higher the sooner the emissions are reduced. Turning to location, a universal price results in developed countries continuing to pollute while buying credits from developing countries. The more desirable outcome is developed countries investing in energy efficiency and greener energy and industrial processes. The third challenge is the now well-identified[12]

---

[11] While the use of GWP (100) has been disputed (Shackley and Wynne, 1997; Shine *et al.*, 2005), it is still used in the IPCC's *Climate Change 2022: Mitigation of Climate Change — Summary for Policymakers*.

[12] The former head of the Australian Government's Emission Reduction Assurance Committee, Professor Andrew MacIntosh, has outlined major integrity issues in four papers. In *Fixing the Integrity Problems with Australia's Carbon Market*, the authors say, "#Earlier this year, we went public with details of serious integrity issues with the ERF, labelling it 'environmental and taxpayer fraud'…The decision to use the word 'fraud' was deliberate and considered. Offset projects are given a financial instrument (a carbon credit) in return for providing a service: the delivery of real and additional greenhouse gas abatement equivalent to one tonne of carbon dioxide for each carbon credit received. Where the credited abatement is not real or is not additional, the service has not been provided. In our view, a process that systematically pays people to provide a service that is not provided is fraudulent. We do not suggest proponents have acted unlawfully. The problem is with the system…not the individual beneficiaries of it." The other papers are *Integrity and the ERF's Human- Induced Regeneration Method: The Measurement Problem Explained, Integrity and the ERF's Human-Induced Regeneration Method: The Additionality Problem Explained,* and *Integrity Problems with the ERF's 2022* Plantation Forestry Method.

potential for corruption in claimed reductions. One does not have to sub-scribe to the entirety of the Marxian analysis[13] to accept the critique of carbon markets.

2.3.2. *Structures, institutions, and actors*

The core structural element on climate change policy remains that it is essential that all countries/economies commit to reducing or eliminating of anthropogenic GHGs, or reduction in GHG sinks, such as forests. The United Nations Framework Convention on Climate Change (UNFCCC) is structural, as it expresses this global commitment (United Nations, 1992). Similarly, the suite of technologies that result in current GHG emissions, the technologies that can replace these with fewer or no emissions, and the technologies that can remove GHGs from the atmosphere, are a family of structural variables.

The third major structural variable is stages of economic develop-ment. Industrialized nations, either collectively the West or the North, and some which are not in either group, such as Australia, have been respon-sible for the vast bulk of the GHGs already in the atmosphere, due essen-tially to industrialization. Linked to the question of global cooperation is that of equity and ensuring developing economies can increase their energy intensity, defined as unit of energy per unit of GPDP.

The institutional element most directly relevant to energy policy under climate change is a consequence of the UNFCCC, that is, the Nationally Determined Contributions requirement under the Paris Agreement (United Nations, 2015). The second relevant institutional ele-ment is the agreement among developed countries to work together on energy security issues arising from the oil shocks of the 1970s. These in turn give rise to the IPCC and the IEA, respectively, the latter having been initially based around the OECD, which itself grew out of the Marshall Plan and primarily consists of the developed Western nations.

Current technology also operates as a specific institution in climate matters, as the socio-technical regime shapes thinking. This is apparent in issues such as the pursuit of inertia through the transition, which ignores

---

[13] The complete Marxian analysis extends to the proposition that the "commodification of climate" has resulted in transnational capital evading emissions reduction opportunities in their home countries while also extracting windfall profits from developing countries without promoting emissions reductions there either.

the fact that on the other side of the transition there is no need for inertia (Geels *et al.*, 2017). The major international actors have already been identified. To these should be added such regional organizations as ASEAN, because of the potential for electricity trading. Other international actors include well-organized environmental advocacy groups such as Greenpeace and the World Wide Fund for Nature (WWF).

How much agency national governments have as actors depends on the nature of the governance structure and the way policy is developed and administered. One objective of the 1980s and 90s reform was reducing political interference in energy (Littlechild, 2006). At the national level, the next most significant actors are those invested in existing energy firms. Because the bulk of energy consumed comes from fossil fuels, either directly or through conversion to electricity, these actors are both heavily resourced and well-engaged. Many are multinational.

The role these firms have played in the global response to climate change is critical to understanding the political economy of CCUS. These firms have argued that climate change is not a problem (Mayer 2016, pp. 198–225; McMullen, 2022), assisted by media being hounded to give equal weight to dissident views and established scientists, as stated by Eugene Linden, a former *Time* environment reporter, "hounded by experts who conflated scientific diffidence with scientific uncertainty, and who wrote outraged letters to the editor when a report didn't include their dissent" (Oreskes and Conway, 2010). Oreskes and Conway (2010) conclude, "Editors evidently succumbed to this pressure, and reporting on climate in the United States became biased toward the skeptics and deniers because of it." One study noted that a third of articles in 10 Australian newspapers during a three month in 2011–2012 did not accept the consensus of anthropogenic climate change (Bacon, 2013).

The interplay between governments, activists, and corporations creates the field on which the political economy of climate change plays out. CCUS is then a family of options on which each of these actors takes a position. Four strategic pathways for CCUS integration can be identified, at large point-source industrial facilities, including power plants, from mobile, i.e., transport, or smaller fixed sources, capture from biomass production, and direct air capture. The first of these was also the first identified pathway and retains most focus. Consequently, its greatest proponent, other than government and inter-governmental agencies, is fossil fuel companies.

3. Competing Frames for CCUS Analysis

There are four identifiable frames for CCUS analysis: technology, markets, environment, and society. The PE factors relating to specific projects can vary. For example, one study found greater acceptance of CCUS in fossil-fuel-rich locations (Kern *et al.*, 2016).

In the Technology Frame, the core issue for CCUS is development. In essence, once the technology is proven, wide-scale adoption will follow. It is exemplified by repeated invocations for more large-scale demonstration projects and government support (de Coninck *et al.*, 2009; Scott *et al.*, 2013). This is the current position of the IEA (IEA, 2020), which also calls for "government leadership and policy support" (IEA, 2021). It also typically underestimates the issues that determine what projects government can and will support (Torvanger and Meadowcroft, 2011).

In the Markets Frame, the focus is on providing CCUS developers with income streams. The simple answer is carbon pricing and markets. The focus of researchers is the level of carbon price necessary to finance the technology (Lau *et al.*, 2021; Scott *et al.*, 2013; Walsh *et al.*, 2014; Wu *et al.*, 2013).

In the Environmental Frame, the core issue is the scope of CCUS application. Recognizing that current progress in reducing GHGs is insufficient to fulfill the Paris Agreement, and that direct air capture may become significant, is uppermost (IEA, 2022; Keith, 2009).

In the Society Frame, the core issue is who benefits from CCUS, and who it affects. While some studies suggest widespread community acceptance of CCUS (van Alphen *et al.*, 2007), others indicate a less positive attitude, albeit with little outright rejection (de Best-Waldhober *et al.*, 2009). One study finds no evidence that support for CCUS currently detracts from renewables support, while noted it was probably too early to detect such effects (Shackley *et al.*, 2009). Another study, spanning four countries found, however, that "it was deemed particularly important by many participants that investments into renewable energy technologies should not suffer because of investment in CCS" (Ashworth *et al.*, 2013). Allen and Chatterton (2013) found greater support for wind, solar, and hydro over other technologies including CCUS and nuclear, with nuclear the least preferable, while Li *et al.* (2014) and de Best-Waldhober *et al.* (2009) found public acceptance of CCUS to be lower than other lower carbon technologies.

L'Orange Seigo *et al.* (2014) observe that CCUS "project developers are typically energy companies. They are the least trusted stakeholder, and even honest communication might not achieve its goal." Similarly, Mabon *et al.* (2013) found that "stakeholders often evaluate the geological storage of carbon dioxide in terms of its relation to their broader world views." Environmental activists, however, have invested in studies of existing CCUS projects. One study found that failed and underperforming projects outnumber successful experiences, and that "using carbon capture as a greenlight to extend the life of fossil fuels power plants is a significant financial and technical risk" (Robertson and Mousavian, 2022).

4. Propositions Concerning CCUS PE

In this section, we construct five propositions on the political economy of CCUS, based on a synthesis of the findings from the above frame-based and political economy analyses.

4.1. *CCUS is too broad*

While CCUS as a single technology descriptor is convenient and descriptive, it is not useful in policy development. It ranges from the first CCS application of extending the life of fossil fuel power plants to the essential eventual atmospheric scrubbing and use. The generation of brown hydrogen as industrial feedstock, though not energy carrier, from natural gas and sequestering CO_2 in gas wells is a useful interim technology.

However, all forms of CCUS acquire the negative sentiment of fossil fuel combustion, both because these projects are failing, either technically or financially, and because, as mentioned above, they are promoted by fossil fuel companies that are mistrusted because of their decades of investment in arguing against anthropogenic climate change. As a consequence, there is a growing negative perception of all forms of CCUS projects, especially among environmental activists.

4.2. *There has been a failure of analysis*

The central role afforded to CCUS in global policy analysis fails to reflect the slow pace of its development. Despite repeated recognition that CCUS is not developing fast enough to fulfill the role expected of it, the IEA and IPCC continue to afford it deference.

Repeated invocations for more investment in demonstrator projects will not close the gap. Policy analysis must instead focus why investment is lacking. One reason is the aforementioned political opposition based on the fossil fuels association. Another is that supporting CCUS is seen as inconsistent with calls for no new fossil fuel projects.

Policy analysis needs to reconsider the role of CCUS in decarbonization, which must follow net-zero scenarios consistent with actual and projected states of these technologies. A consequence of unrealistic scenarios is that policymakers at national levels and industry stakeholders alike incorporate them in their own assumptions.

4.3. *Markets are problematic*

There is widespread perception that market reforms to energy supply are part of our current energy problems (Varoufakis, 2022). Consumers have not regarded energy market reform as a success, which they measure in price. The real price increases observed in Australia this century have are similar to those in most other markets.

Observed average residential and total market prices (AU)

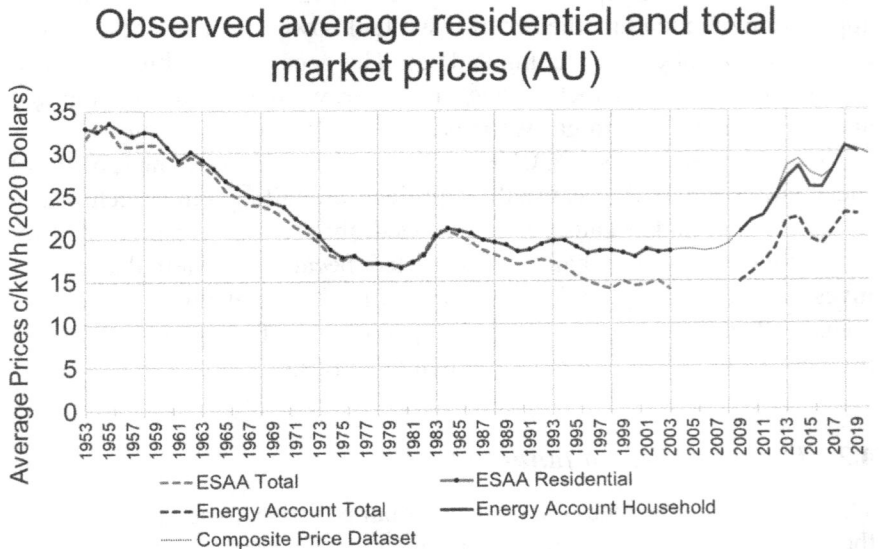

Legend:
- - - ESAA Total
- - - Energy Account Total
----- Composite Price Dataset
——•— ESAA Residential
——— Energy Account Household

Y-axis: Average Prices c/kWh (2020 Dollars)

Apart from the more general philosophical stance of rejecting "marketizing the environment," citizens are wary of the suggestion that a market for carbon solves all problems, despite the necessity of such a market to most CCUS projects. There is a further perception that because large corporations, especially the aforementioned fossil fuel companies, back carbon prices then they must benefit. Finally, carbon prices are seen as unjust, because they increase per unit energy costs, tantamount to regressive taxation on low-income households, which can spend up to 10 percent of their budgets on energy. While some carbon price implementations, such as Australia's short-lived CPRS, are accompanied by social adjustment packages, these are expensive to implement for governments that may deriving no new revenue from carbon pricing.

4.4. *Proponents are a problem*

We have already noted in propositions 1 and 3 that CCUS proponents are part of the problem. There is deep distrust of corporations, especially fossil fuel companies, supported by the long record of corporations abusing science to further corporate goals, i.e., tobacco, asbestos (Mayer, 2016; Oreskes and Conway, 2010), including the fact that companies also did so respecting climate change (Mayer, 2019).

It does not help that these corporations act as if society owes them a living. This behavior is grounded in two corporate beliefs. The first is a misinterpretation of the concept of "regulatory taking," which is generally an expression of property rights and a protection in most liberal democracies against unjust takings. This protection does not extend to regulatory risk that firms can and should contemplate, however. The second is the idea that economic success depends on corporate success, famously stated by the CEO of General Motors as, "What's good for GM is good for the country."[14] This is unfounded because overall corporate success does not require all corporations to succeed, and, especially in essential services, government ownership remains a viable alternative.

---

[14] The full correct quote is, "For years I thought that what was good for our country was good for General Motors and vice versa. The difference did not exist. Our company is too big. It goes with the welfare of the country." Congressional confirmation hearing on appointment of Wilson as Secretary of Defense, January 15–16, 1953. https://www.google.com.au/books/edition/Nominations/jtpLAQAAIAAJ?hl=en&gbpv=0.

Finally, communities are increasingly concerned about the outsized influence of lobbying and influence networks. This extends from the weight of submissions that industry can make in formal processes through to corporate donations to political campaigns. One historic study of the nexus between energy and economic growth concluded that "states must prevent vested interests from blocking structural change. States that are unable to do this will get locked into yesterday's technologies, industries and energy systems, effectively consigning themselves to stagnation and decline" (Moe, 2010).

4.5. *Sunk cost "recovery" is bad strategy*

It is a first principle of economics that historic single-purpose investments that have already been made, that is sunk costs, should not be considered in analyses of future action (Frank *et al.*, 2007, p. 10; Mankiw, 2007, p. 297). While McAfee *et al.* (2010) note that it may be rational to consider sunk costs due to informational content, reputational concerns, or financial and time constraints, these are mostly in relation to making further investments to complete projects underway. In general, however, investments in fossil fuel generators cannot be salvaged by CCUS, and investing in it to extend their life is thus bad policy, made worse if further investment in fossil fuel plants is based on an assumption that somewhere in the future CCUS will appear that itself requires investments of unknown size and subsidy, i.e., carbon price, of unknown form.

5. Conclusion: CCUS Best Practices

In this section, we discuss important policy lessons pertaining to CCUS best practices, followed by closing remarks.

5.1. *Avoid strategy lock-in*

Further investment in fossil fuel generation, and delaying electrification of other energy applications, results in strategy lock-in, where a nation reaches a point where only CCUS can meet emissions reduction goals. At present, past fossil fuel investments need not determine the future. That means ignoring CCUS boosterism from the IEA and IPCC, which, as noted, continue to place CCUS on the decarbonization pathway, despite its not yet achieving the requisite maturity.

Technology forecasting is fraught with danger. An Australian physics professor wrote in a 1953 paper that "there is little or no hope that power from wind, tides and waterfalls will take over any large fraction of our power requirements. The utilization of solar energy is not as easy as many people have been led to believe and is not likely to play a major role in supplying power for many hundreds of years" (Messel, 1953). While the silicon solar cell was developed at Bell Labs the following year, the *New York Times* erred in the opposite direction, forecasting "the harnessing of the almost limitless energy of the sun for the uses of civilization" (Anon., 1954).

Similarly, the liquefied natural gas industry was also slow to develop, with the first plant built in 1912, the first working storage in 1941, the first ship in 1959, and the next three ships a decade later. This is easier said than done, given the powerful global forces supporting CCUS. Corporations are as guilty as the IEA, if not more so, in unwarranted faith in CCUS to be the savior of fossil fuel plants in a net-zero scenario. However, if they seriously believe in its potential, they should be entirely financing its development.

5.2. *Actively work on alternatives*

The narrative that "CCUS is important, and we need people to be aware of CCUS to promote CCUS policies" is unhelpful. Policymakers need alternatives to avoid strategy lock-in. Rather than starting with the assumption of CCUS, plan using reframing scenarios. We noted earlier that policy controversies are grounded in participants approaching issues from their respective frames. We identified CCUS analysis frames, noting the conflict between the technological and market frames and the environmental and society frames. Resolution is not found in endless additional debates between actors who are firmly grounded in their perception of what matters.

Thankfully, there is an alternative. A book called *Strategic Reframing* outlines the Oxford Scenario Planning Approach (Ramirez and Wilkinson, 2016), a method to reframe Turbulence, Unpredictable Uncertainty, Novelty, and Ambiguity, or TUNA issues, which describes all of the wicked problems of policy controversy. There may be better options than CCUS to abate fossil fuel power plant impacts. It might be better to pursue more renewables for electricity and use carbon sinks to earn export income. It may also be better to pursue industrialization strategies for participation in renewables. The latter becomes increasingly important as countries look to diversify supply chains away from China. The point, however, is not to identify alternative strategies, but that countries should use the aforementioned scenario-planning approaches to give themselves options.

CCUS political economy analysis finds that reliance on its eventual development to enable continued use of fossil fuel power plants is a strategy with a high risk of failure, as attested by the IEA and IPCC's own acceptance that development is behind where they needed it for their earlier net-zero scenarios. Causes include the extent of the policy controversy surrounding CCUS with a growing gap between technology and market frames and environmental and society frames.

One risk is that other essential CCUS development for other applications may also be delayed. Another is that countries may fail to achieve their emissions reduction targets through unrealized reliance on CCUS. Policymakers must utilize available tools to reframe policy controversies and look to a wider set of net-zero pathways.

References

Akintan, O., Jewitt, S., and Clifford, M. (2018). Culture, tradition, and taboo: Understanding the social shaping of fuel choices and cooking practices in Nigeria. *Energy Research & Social Science*, *40*, 14–22.

Allen, P. and Chatterton, T. (2013). Carbon reduction scenarios for 2050: An explorative analysis of public preferences. *Energy Policy*, *63*, 796–808.

Alt, J. E. and Chrystal, K. A. (1983). *Political Economics* (Vol. 2). California: University of California Press.

Anon. (1954). Vast Power of the Sun is Trapped by Battery Using Sand Ingredient. *New York Times*, 26 April 1954, 1, 11.

APSC. (2007). *Tackling Wicked Problems: A Public Policy Perspective*, Contemporary government challenges (Australian Public Service Commission), Australian Public Service Commission, [Canberra].

Arnsperger, C. and Varoufakis, Y. (2006). What is neoclassical economics? The three axioms responsible for its theoretical oeuvre, practical irrelevance and, thus, discursive power. *Panoeconomicus*, *53*(1), 5–18.

Ashworth, P., Einsiedel, E., Howell, R., Brunsting, S., Boughen, N., Boyd, A., Shackley, S., Van Bree, B., Jeanneret, T., Stenner, K., Medlock, J., Mabon, L., Feenstra, C. F. J., and Hekkenberg, M. (2013). Public preferences to CCS: How does it change across countries? *Energy Procedia*, *37*, 7410–7418.

Averchenkova, A. and Nachmany, M. (2017). Institutional aspects of climate legislation. In A. Averchenkova, S. Fankhauser and M. Nachmany (eds.). *Trends in Climate Change Legislation*. Cheltenham, UK: Edward Elgar Publishing Limited.

Bacon, W. (2013). *Climate Science in Australian Newspapers. Sceptical Climate Part 2: Climate Science in Australian Newspapers*. The Australian Centre for Independent Journalism, Sydney, Australia, https://www.uts.edu.au/sites/default/files/Sceptical-Climate-Part-2-Climate-Science-in-Australian-Newspapers.pdf.

Bardi, U. (2009). Peak oil: The four stages of a new idea. *Energy*, *34*(3), 323–326.

Bardi, U. (2019). Peak oil, 20 years later: Failed prediction or useful insight. *Energy Research & Social Science*, *48*, 257–261.

Barrera-Santana, J., Marrero, G., and Ramos-Real, F. (2022). Energy efficiency and energy governance: A stochastic frontier analysis approach. *The Energy Journal*, *43*(6), 243–283.

Bate, R. (2001). *The Political Economy of Climate Change Science: A Discernible Human Influence on Climate Documents?*, Vol. 1, IEA environment briefing, IEA Environment Unit, London.

Bellantuono, G. (2016). The comparative law and economics of energy markets. In T. Eisenberg and G. B. Ramello (eds.). *Comparative Law and Economics*. London: Edward Elgar Publishing, 236–261.

Bharadwaj, B., Pullar, D., To, L. S. and Leary, J. (2021). Why firewood? Exploring the co-benefits, socio-ecological interactions and indigenous knowledge surrounding cooking practice in rural Nepal. *Energy Research & Social Science*, *75*, 101932.

Boudewijns, E. A., Trucchi, M., van der Kleij, R. M. J. J., Vermond, D., Hoffman, C. M., Chavannes, N. H., van Schayck, O. C. P., Kirenga, B. and Brakema, E. A. (2022). Facilitators and barriers to the implementation of improved solid fuel cookstoves and clean fuels in low-income and middle-income countries: an umbrella review. *The Lancet Planetary Health*, *6*(7), e601–e12.

Brown, M. A. (2001). Market failures and barriers as a basis for clean energy policies. *Energy Policy*, *29*(1)4, 1197–207.

Büscher, C., Schippl, J., and Sumpf, P. (2018). *Energy as a Sociotechnical Problem: An Interdisciplinary Perspective on Control, Change, and Action in Energy Transitions*. Routledge.

Chapman, I. (2014). The end of peak oil? Why this topic is still relevant despite recent denials. *Energy Policy*, *64*, 93–101.

Chester, L. (2010). Conceptualising energy security and making explicit its poly-semic nature. *Energy Policy*, *38*(2), 887–895.

Churchman, C. W. (1967). Guest Editorial "Wicked Problems". *Management Science, 14*(4), B-141–B-6.

Clapp, J. and Dauvergne, P. (2011). *Paths to a Green World, Second Edition: The Political Economy of the Global Environment*. Cambridge: MIT Press.

Coase, R. H. (1959). The federal commissions telecommunications. *Journal of Law and Economics, 3*, 1–44.

Coase, R. H. (1960). The problem of social cost. *Journal of Law and Economics,* 3, 1.

Colander, D. (2000). The death of neoclassical economics. *Journal of the History of Economic Thought, 22*(2), 127–143.

Collinson, S. (2003). *Power, Livelihoods and Conflict: Case Studies in Political Economy Analysis for Humanitarian Action*. London: Humanitarian Policy Group, Overseas Development Institute.

Correljé, A. and De Vries, L. (2008). Hybrid electricity markets: The problem of explaining different patterns of restructuring. In F. Sioshansi (ed.). *Competitive Electricity Markets*. Amsterdam: Elsevier, 65–93.

de Best-Waldhober, M., Daamen, D., Ramirez Ramirez, A., Faaij, A., Hendriks, C., and de Visser, E. (2009). Informed public opinions on CCS in comparison to other mitigation options. *Energy Procedia, 1*(1), 4795–4802.

de Coninck, H., Flach, T., Curnow, P., Richardson, P., Anderson, J., Shackley, S., Sigurthorsson, G., and Reiner, D. (2009). The acceptability of CO_2 capture and storage (CCS) in Europe: An assessment of the key determining factors: Part 1. Scientific, technical and economic dimensions. *International Journal of Greenhouse Gas Control, 3*(3), 333–343.

DeCanio, S. J. (1998). The efficiency paradox: bureaucratic and organizational barriers to profitable energy-saving investments. *Energy Policy, 26*(5), 441–454.

DFAT. (2016), *Political Economy Analysis: Guidance Note, Australia*. Canberra: Department of Foreign Affairs and Trade.

Dowd, D. (2004), *Capitalism and Its Economics: A Critical History* (2nd edn.). Pluto Press.

Frank, R. H., Jennings, S., and Bernanke, B. (2007). *Principles of Microeconomics* (1st Australasian edn.). North Ryde, NSW: McGraw-Hill Australia.

Frieden, J. (2020). The political economy of economic policy: We should pay closer attention to the interactions between politics, economics, and other realms. *Finance & Development, 57*(002).

Friedman, M. (1953). The methodology of positive economics. In *Essays in Positive Economics*, Chicago, IL: University of Chicago Press.

Fukuyama, F. (2011). *The Origins of Political Order: From Prehuman Times to the French Revolution*. London: Profile Books.

Fullbrook, E. E. (2008). *Pluralist Economics*. London: Zed Books.

Geels, F. W. (2002). Technological transitions as evolutionary reconfiguration processes: A multi-level perspective and a case-study. *Research Policy, 31*(8), 1257–1274.

Geels, F. W., Sovacool, B. K., Schwanen, T., and Sorrell, S. (2017). The socio-technical dynamics of low-carbon transitions. *Joule, 1*(3), 463–479.

Glachant, J.-M. and Perez, Y. (2011). The liberalization of electricity markets. In *International Handbook of Network Industries*. London: Edward Elgar Publishing.

Graaf, T. V. d, Sovacool, B. K., Ghosh, A., Kern, F., and Klare, M. T. (2016). States, markets, and institutions: Integrating international political economy and global energy politics. In T. V. d Graaf, B. K. Sovacool, A. Ghosh, F. Kern and M. T. Klare (eds.). *The Palgrave Handbook of the International Political Economy of Energy*. Berlin: Springer, 3–44.

Graaf, T. V. d and Zelli, F. (2016). Actors, institutions and frames in global energy politics. In *The Palgrave Handbook of the International Political Economy of Energy*. Berlin: Springer, 47–71.

Grusin, R. (2015). *The Nonhuman Turn*. Minnesota: University of Minnesota Press.

Harvey, J. T. (2020). *Contending Perspectives in Economics: A Guide to Contemporary Schools of Thought*. Cheltenham: Edward Elgar.

Havyatt, D. (2022). History of Electricity Reform in Australia. In G. Roger (ed.), *On the Grid: Australian Electricity in Transition*. Melbourne: Monash University Publishing. pp. 1–44.

Heilbroner, R. L. (1970). On the possibility of a political economics. *Journal of Economic Issues, 4*(4), 1–22.

IEA. (2020). *CCUS in Clean Energy Transitions*. Paris: IEA.

IEA. (2021). *Carbon Capture, Utilisation and Storage: The Opportunity in Southeast Asia*. Paris: IEA.

IEA. (2022). *Direct Air Capture: A Key Technology for Net-zero*. Paris: IEA.

Jaffe, A. B. and Stavins, R. N. (1994). The energy-efficiency gap: What does it mean? *Energy Policy, 22*(10), 804–810.

Jakob, M., Flachsland, C., Steckel, J. C., and Urpelainen, J. (2020). Actors, objectives, context: A framework of the political economy of energy and climate policy applied to India, Indonesia, and Vietnam. *Energy Research and Social Science, 70*, 101775.

Joskow, P. L. (2006). Introduction to electricity sector liberalization: Lessons learned from cross country studies. In F. Sioshansi and W. Pfaffenberger (eds.). *Electricity Market Reform: An International Perspective*. Amsterdam, The Netherlands: Elsevier Science, 1–32.

Kay, J. (2007). The failure of market failure', *Prospect*, 1 August 2007.

Keen, S. (2011). *Debunking Economics: The Naked Emperor Dethroned?* (2nd edn.). London, UK: Zed Books.

Keith, D. W. (2009). Why capture CO_2 from the atmosphere? *Science, 325*(5948), 1654–1655.

Kern, F., Gaede, J., Meadowcroft, J., and Watson, J. (2016). The political economy of carbon capture and storage: An analysis of two demonstration projects. *Technological Forecasting and Social Change, 102*, 250–260.

Kessides, I. N. (2004). *Reforming Infrastructure: Privatization, Regulation, and Competition*. Washington, DC: World Bank Publications.

Kocka, J. (2016). *Capitalism: A Short History*. Princeton, NJ: Princeton University Press.

L'Orange Seigo, S., Dohle, S., and Siegrist, M. (2014). Public perception of carbon capture and storage (CCS): A review. *Renewable and Sustainable Energy Reviews*, *38*, 848–863.

Lau, H. C., Ramakrishna, S., Zhang, K., and Radhamani, A. V. (2021). The role of carbon capture and storage in the energy transition. *Energy & Fuels*, *35*(9), 7364–7386.

Lawson, T. (2006). The nature of heterodox economics. *Cambridge Journal of Economics*, *30*(4), 483–505.

Lawson, T. (2013). What is this "school" called neoclassical economics? *Cambridge Journal of Economics*, *37*(5), 947–983.

Lawson, T. (2017). What is wrong with modern economics, and why does it stay wrong? *Journal of Australian Political Economy*, 80, 26–42.

Li, Q., Liu, L.-C., Chen, Z.-A., Zhang, X., Jia, L., and Liu, G. (2014). A survey of public perception of CCUS in China. *Energy Procedia*, *63*, 7019–7023.

Liao, H., Chen, T., Tang, X. and Wu, J. (2019). Fuel choices for cooking in China: Analysis based on multinomial logit model. *Journal of Cleaner Production*, *225*, 104–111.

Littlechild, S. C. (1979). *The Fallacy of the Mixed Economy: An "Austrian" Critique of Conventional Economics and Government Policy*. California: Cato Institute.

Littlechild, S. C. (2006). Preface: Electricity Market Reform. In F. Sioshansi and W. Pfaffenberger (eds.). *Electricity Market Reform: An International Perspective*. Amsterdam, The Netherlands: Elsevier Science, xvii–xxix.

Lutzenhiser, L. and Shove, E. (1999). Contracting knowledge: The organizational limits to interdisciplinary energy efficiency research and development in the US and the UK. *Energy Policy*, *27*(4), 217–227.

Mabon, L., Vercelli, S., Shackley, S., Anderlucci, J., Battisti, N., Franzese, C., and Boot, K. (2013). 'Tell me what you think about the geological storage of carbon dioxide': Towards a fuller understanding of public perceptions of CCS. *Energy Procedia*, *37*, 7444–7453.

MacKenzie, D. (2009). Making things the same: Gases, emission rights and the politics of carbon markets. *Accounting, Organizations and Society*, *34*(3–4), 440–455.

Malakar, Y., Greig, C., and van de Fliert, E. (2018). Structure, agency and capabilities: Conceptualising inertia in solid fuel-based cooking practices. *Energy Research & Social Science*, *40*, 45–53.

Mallin, S. (2011). Market Fetishism, or,"Thinking like an Economist". *Rethinking Marxism*, *23*(1), 41–49.

Mankiw, N. G. (2007). *Principles of Microeconomics* (4th edn.). Ohio: Thomson/South-Western, Mason. Marechal, K. (2012). *The Economics of Climate Change and the Change of Climate in Economics.* London: Routledge.

Masera, O. R., Saatkamp, B. D., and Kammen, D. M. (2000). From linear fuel switching to multiple cooking strategies: A critique and alternative to the energy ladder model. *World Development, 28*(12), 2083–2103.

Mayer, J. (2016). *Dark Money: The Hidden History of the Billionaires behind the Rise of the Radical Right* (1st edn.). New York: Doubleday.

Mayer, J. (2019). "Kochland" examines the Koch Brothers' early, crucial role in climate-change denial. *New Yorker*, 13 August 2019.

McAfee, R. P., Mialon, H. M. and Mialon, S. H. (2010). Do sunk costs matter? *Economic Inquiry, 48*(2), 323–336.

McMullen, J. (2022). *The Audacious PR Plot that Seeded Doubt About Climate Change*, viewed 4 Apr 2024 2022, https://www-bbc-co-uk.cdn.ampproject. org/c/s/www.bbc.co.uk/news/science-environment-62225696.amp.

Messel, H. (1953). Nuclear power for Australian industry. *The Australian Quarterly, 25*(4), 7–12.

Micklethwait, J. and Wooldridge, A. (2005). *The Company: A Short History of a Revolutionary Idea*, 12. London, UK: Modern Library.

Mirowski, P. (1991). *More Heat than Light: Economics as Social Physics, Physics as Nature's Economics.* Cambridge: Cambridge University Press.

Moe, E. (2010). Energy, industry and politics: Energy, vested interests, and long-term economic growth and development. *Energy, 35*(4), 1730–1740.

Nepal, R. and Jamasb, T. (2012). Reforming the power sector in transition: Do institutions matter? *Energy Economics, 34*(5), 1675–1682.

Oreskes, N. and Conway, E. M. (2010). *Merchants of Doubt: How a Handful of Scientists Obscured the Truth on Issues from Tobacco Smoke to Global Warming* (1st US edn.). New York: Bloomsbury Press.

Quesada, A. U. (1989). Greenhouse economics: Global resources and the political economy of climate change. *Environmental Policy & Law Journal, 19*, 154.

Quiggin, J. (2010). *Zombie Economics: How Dead Ideas Still Walk among Us.* Princeton, NJ: Princeton University Press.

Ramirez, R. and Wilkinson, A. (2016), *Strategic Reframing: The Oxford Scenario Planning Approach.* Oxford: Oxford University Press.

Rittel, H. W. and Webber, M. M. (1973). Dilemmas in a general theory of planning. *Policy Sciences, 4*(2), 155–169.

Robertson, B. and Mousavian, M. (2022). *The Carbon Capture Crux.* Lakewood, Ohio, USA: Institute for Energy Economics and Financial Analysis. https://ieefa.org/resources/carbon-capture-crux-lessons-learned.

Rothenberg, L. (2012). The Political Economy of Climate Change. In *Responding to Climate Change.* Cheltenham: Edward Elgar Publishing.

Rudolph, S. and Aydos, E. (2021). *Carbon Markets Around the Globe: Sustainability and Political Feasibility*. Cheltenham: Edward Elgar Publishing Limited.

Schabas, M. (1993). What's so wrong with physics envy? *History of Political Economy*, 25(1), 45–53.

Schön, D. A. and Rein, M. (1994). *Frame Reflection: Toward the Resolution of Intractable Policy Controversies*. New York: Basic Books.

Scott, V., Gilfillan, S., Markusson, N., Chalmers, H., and Haszeldine, R. S. (2013). Last chance for carbon capture and storage. *Nature Climate Change*, 3(2), 105–111.

Scrase, I., Wang, T., MacKerron, G., McGowan, F., and Sorrell, S. (2009). Introduction: Climate policy is energy policy. In I. Scrase & G. MacKerron (eds.), *Energy for the Future*. Berlin: Springer, 3–19.

Shackley, S., Reiner, D., Upham, P., de Coninck, H., Sigurthorsson, G., and Anderson, J. (2009). The acceptability of CO_2 capture and storage (CCS) in Europe: An assessment of the key determining factors: Part 2. The social acceptability of CCS and the wider impacts and repercussions of its implementation. *International Journal of Greenhouse Gas Control*, 3(3), 344–356.

Shackley, S. and Wynne, B. (1997). Global Warming Potentials: ambiguity or precision as an aid to policy? *Climate Research*, 8(2), 89–106.

Shine, K. P., Fuglestvedt, J. S., Hailemariam, K., and Stuber, N. (2005). Alternatives to the global warming potential for comparing climate impacts of emissions of greenhouse gases. *Climatic Change*, 68(3), 281–302.

Simon, H. A. (1997) (1945). *Administrative Behaviour*. London: Free Press.

Sioshansi, F. P. and Pfaffenberger, W. (2006). *Electricity Market Reform: An International Perspective*. Amsterdam: Elsevier.

Skaburskis, A. (2008). The origin of "wicked problems". *Planning Theory & Practice*, 9(2), 277–280.

Sovacool, B. K. (2011). *The Routledge Handbook of Energy Security*. Abingdon: Routledge.

Sovacool, B. K. (2014). What are we doing here? Analyzing fifteen years of energy scholarship and proposing a social science research agenda. *Energy Research & Social Science*, 1, 1–29.

Sovacool, B. K. and Brown, M. A. (2015). Deconstructing facts and frames in energy research: Maxims for evaluating contentious problems. *Energy Policy*, 86, 36–42.

Spreng, D. (2014). Transdisciplinary energy research–reflecting the context. *Energy Research & Social Science*, 1, 65–73.

Stilwell, F. (2011). Marketising the environment. *The Journal of Australian Political Economy*, 68, 108–127.

Stilwell, F. (2012). *Political Economy: The Contest of Economic Ideas* (3rd edn.). Oxford: Oxford University Press.

Torvanger, A. and Meadowcroft, J. (2011). The political economy of technology support: Making decisions about carbon capture and storage and low carbon energy technologies. *Global Environmental Change*, *21*(2), 303–312.

Tversky, A. and Kahneman, D. (1974). Judgment under uncertainty: Heuristics and biases. *Science*, *185*(4157), 1124–1131.

Tversky, A. and Kahneman, D. (1985). The framing of decisions and the psychology of choice. In *Behavioral Decision Making*. Berlin: Springer, 25–41.

United Nations (1992). *United Nations Framework Convention on Climate Change*, United Nations.

United Nations (2015). *Paris Agreement*. United Nations. New York: Framework Convention on Climate Change, https://unfccc.int/sites/default/files/english_paris_agreement.pdf.

Valentine, S. V. (2010). The fuzzy nature of energy security. In B. K. Sovacool (ed.), *The Routledge Handbook of Energy Security*. Abingdon: Routledge, 74–91.

van Alphen, K., van Voorst, Q., Hekkert, M. P. and Smits, R. E. H. M. (2007). Societal acceptance of carbon capture and storage technologies. *Energy Policy*, *35*(8), 4368–4380.

Van den Bergh, J. C. (2001). Ecological economics: Themes, approaches, and differences with environmental economics. *Regional Environmental Change*, *2*(1), 13–23.

Varoufakis, Y. (2022). Time to blow up electricity markets. https://www.project-syndicate.org/commentary/marginal-cost-pricing-for-electricity-disastrous-in-europe-by-yanis-varoufakis-2022-08.

Vickers, G. (1965). *The Art of Judgment; A Study of Policy-Making*. New York: Basic Books.

Walsh, D. M., O'Sullivan, K., Lee, W. T. and Devine, M. T. (2014). When to invest in carbon capture and storage technology: A mathematical model. *Energy Economics*, *42*, 219–225.

Weintraub, E. R. (2002). How economics became a mathematical science. In *How Economics Became a Mathematical Science*. North Carolina: Duke University Press.

Williamson, O. E. (1995). The institutions and governance of economic development and reform. In: *Proceedings of the World Bank Annual Conference on Development Economics 1994, World Bank*, pp. 171–197.

Williams, J. H. and Ghanadan, R. (2006). Electricity reform in developing and transition countries: A reappraisal. *Energy*, *31*(6–7), 815–844.

Wu, N., Parsons, J. E. and Polenske, K. R. (2013). The impact of future carbon prices on CCS investment for power generation in China. *Energy Policy*, *54*, 160–172.

Xie, F. and Cheng, H. (2021). The political economy of climate change: The impasse and way out. *Economic and Political Studies*, *9*(3), 315–335.

Index